THE ELEGANT SOLUTION: HOW NATURE PRESERVES ITS OWN SECRETS

I can't change the past but I'm damn sure going to change the future - Justin Sirotin

Copyright November 14th, 2024 (Happy birthday Mom)

About the Author

My journey to theoretical physics is as unconventional as my groundbreaking theories. My path from CNN producer to prison inmate to AI entrepreneur might seem an unlikely route to developing a unified theory of physics, but it's precisely this unique trajectory that enabled me to see patterns others missed.

My neurodivergent mind naturally sought out underlying patterns in everything from human behavior to the fundamental fabric of reality. My early success at CNN demonstrated my ability to piece together complex narratives from seemingly unrelated fragments of information – a skill that would later prove invaluable in theoretical physics.

It was during my time in prison, stripped of all external distractions, that I began to see connections between quantum mechanics, gravitational theory, and the concept of emergence. The structured environment of incarceration, combined with my pattern-recognition abilities, led me to develop unique insights into the nature of reality itself. My "Seven Pillars of Creation" philosophy, initially developed as a framework for personal transformation, unexpectedly provided a metaphorical foundation for understanding the layered nature of physical reality.

The Spinor-Mediated Universal Transition (SMUT) model represents the culmination of this unconventional journey – a theory that challenges traditional approaches to unifying quantum mechanics and general relativity. Like me, SMUT refuses to be constrained by conventional boundaries, offering a radical new perspective on the nature of reality itself.

Today, I continues to bridge worlds – between classical and quantum physics, between AI and human consciousness, between conventional wisdom and revolutionary insight. My work stands as testament to the power of thinking differently, and to the profound insights that can emerge from the most unexpected places.

CHAPTER 1:

Two Rule Books for One Universe

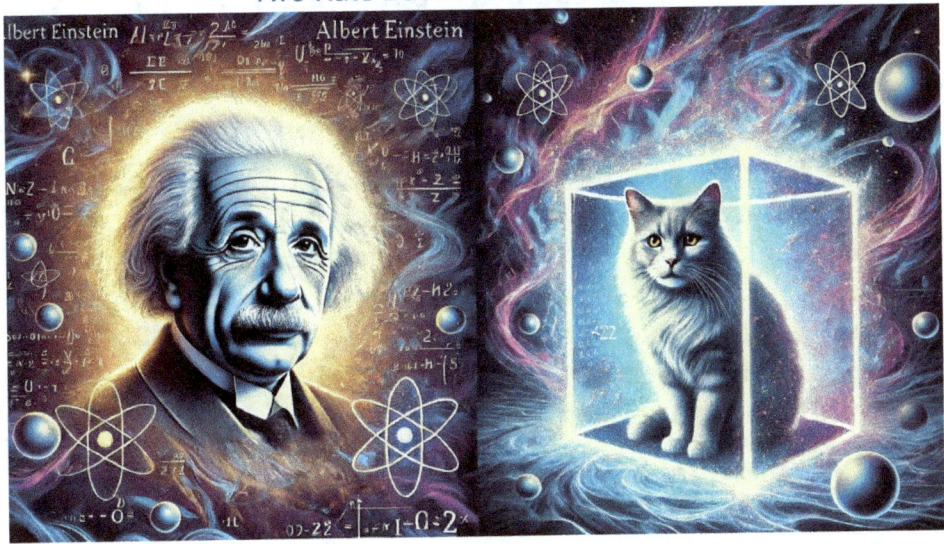

Picture yourself holding two rule books. In your right hand is a thick volume titled "General Relativity" - Einstein's masterwork that explains how gravity shapes our cosmos, how massive objects warp the fabric of spacetime, and why planets orbit stars in elegant elliptical paths. In your left hand is an equally weighty tome called "Quantum Mechanics" - the bizarre but incredibly accurate description of the atomic realm, where particles exist in multiple places at once and seem to communicate instantaneously across vast distances.

These two rule books have served physics extraordinarily well. General relativity precisely predicts the bending of starlight around the Sun and the rhythmic dance of binary pulsars. Quantum mechanics gives us lasers, transistors, and explains why atoms don't collapse. Both theories work brilliantly in their own domains.

But here's the problem: these rule books fundamentally disagree with each other. It's as if one book was written in English and the other in Klingon, with no reliable translation between them. When we try to apply quantum mechanics to gravity, or general relativity to atomic phenomena, our equations explode with meaningless infinities. Nature, clearly, doesn't have this problem - the universe seamlessly combines the very large and the very small. But our best theories can't seem to do the same.

This is where the Spinor-Mediated Universal Transition model - SMUT for short - enters our story. Instead of trying to force these two rule books to play nice, SMUT suggests a radical alternative: what if there's a deeper principle at work? What if the apparent conflict between quantum mechanics and general relativity is telling us something profound about how the universe maintains continuity across different scales?

Think of it like this: imagine you're looking at a beautiful tapestry. Up close, you see individual threads woven in intricate patterns. Step back, and those same threads merge into a grand image. Both views are valid and true, but they seem totally different. What SMUT proposes is that there's a way to understand how those views connect - not by forcing one perspective to work at the wrong scale, but by understanding the principles that allow smooth transitions between scales.

At the heart of this new framework are two key ideas: spinors and torsion. Don't worry if these terms sound alien - we'll explore them properly in the next chapter. For now, think of spinors as nature's messengers, carrying information across different scales of the universe, and torsion as a special kind of twist in the fabric of spacetime that allows for smooth transitions between those scales.

This isn't just abstract theory - SMUT makes specific, testable predictions about everything from the echoes of gravitational waves to the patterns we might find in the cosmic microwave background radiation. It suggests solutions to some of physics' biggest mysteries: the nature of dark matter, the fate of information that falls into black holes, and even why our universe seems to be made almost entirely of matter rather than antimatter.

But perhaps most importantly, SMUT offers a new way of thinking about our universe - not as a collection of separate domains governed by different rules, but as a deeply interconnected system where preservation and continuity are fundamental principles. It hints at a cosmos more strange and beautiful than we had imagined, where the very large and very small are linked in subtle but profound ways.

As we embark on this journey to understand SMUT, we'll encounter ideas that might seem radical or even impossible at first glance. That's okay - some of the most important advances in physics initially seemed outlandish. Remember that when Einstein first suggested that time itself could stretch and squeeze like rubber, or when quantum mechanics proposed that particles could exist in multiple places simultaneously, these ideas seemed just as strange.

Keep an open mind, and get ready to explore a new way of thinking about the universe - one that might just solve some of our deepest cosmic mysteries.

CHAPTER 2:

The Universe's Secret Messengers

Here's a fun experiment: pick up a coffee cup and rotate it 360 degrees. It's back where it started, right? Now imagine something strange - an object that needs to be rotated *twice* around, a full 720 degrees, before it truly returns to its original state. Welcome to the weird world of spinors.

Spinors are like the universe's secret messengers, and they're stranger than anything we've dreamed up in science fiction. While normal objects in our everyday world complete their rotation in one 360-degree turn, spinors are different. They have to go around twice, completing a 720-degree rotation, to return to their starting point. If this seems impossible to visualize, you're in good company - even Richard Feynman once demonstrated this peculiar property with a filled coffee cup attached to his belt by rubber bands, performing what became known as the "Feynman plate trick."

But why should we care about these mathematical oddities?

Because spinors might just be the key to understanding how our universe maintains its continuity across vastly different scales. Think of them as nature's diplomats, capable of carrying information between the quantum realm and the cosmic scale. They're like universal translators in a sci-fi story, but instead of translating between alien languages, they translate between different layers of reality.

This is where torsion enters our story. If spinors are the messengers, torsion is the medium they travel through. But what exactly is torsion? Imagine spacetime not as a simple rubber sheet (the usual analogy for Einstein's gravity) but as a fabric that can both bend *and* twist. It's like the difference between pushing down on a trampoline and wringing out a wet towel - both change the shape of the material, but in fundamentally different ways.

When Einstein described gravity, he showed us how mass curves spacetime - creating the familiar force that keeps planets in orbit and our feet on the ground. But SMUT suggests there's more to the story. The addition of torsion allows spacetime to twist in ways that create smooth transitions between different scales of the universe. It's as if the fabric of reality itself can adapt and flex to maintain continuity between the quantum and cosmic realms.

Together, spinors and torsion form a dynamic duo. The spinors carry information across scales, while torsion provides the twisted pathways they travel through. It's a bit like a cosmic postal service, where the spinors are the mail carriers and torsion creates the roads they travel on. But these roads aren't just flat highways - they're more like Möbius strips and Klein bottles, mathematical objects with strange topological properties that allow for smooth transitions between different dimensions.

This might all sound abstract, but the implications are profound. In traditional physics, we hit a wall when trying to connect the quantum and cosmic scales - our equations break down, producing meaningless infinities. But spinors and torsion working together might provide a way around this problem. They could explain how information is preserved when it falls into a black hole, how quantum effects influence cosmic structures, and even why the universe appears to be filled with mysterious dark matter.

Think of it this way: imagine you're watching a skilled acrobat perform. From a distance, their movements appear smooth and continuous. Up close, you might see the individual muscle contractions and micro-adjustments that make those movements possible. Both perspectives are valid, but they seem very different. Spinors and torsion are like the underlying principles that connect these perspectives - they explain how the small-scale adjustments add up to create the smooth, large-scale movement.

The beauty of this framework is that it's not just mathematical gymnastics - it makes specific, testable predictions. If SMUT is correct, we should see particular patterns in gravitational waves, specific signatures in the cosmic microwave background, and unique relationships between the rotation of galaxies at different scales. It's as if the universe has left us a trail of breadcrumbs, encoded in these phenomena, that could lead us to a deeper understanding of reality.

As we move forward, we'll explore these predictions in detail and see how they might help us solve some of the biggest mysteries in modern physics. But for now, let's appreciate the elegant simplicity of this idea: that the universe maintains its continuity through these strange messengers - spinors - traveling along twisted paths in the fabric of spacetime itself.

Remember that coffee cup rotation we started with? It turns out that simple demonstration might be pointing us toward one of the deepest principles in nature - that the universe has built-in mechanisms for preserving information and maintaining continuity across all scales. Sometimes the most profound insights come from the simplest observations, if we just know how to look at them in the right way.

CHAPTER 3:

Black Holes: Nature's Most Aggressive Data Hoarders

Imagine the universe's strictest librarian. They never lose a book, never misfile a document, and maintain perfect records of everything that has ever happened. Now imagine that same librarian discovering that some of their precious information is vanishing into cosmic paper shredders scattered throughout space. That's essentially the crisis physicists faced when studying black holes - and boy, were they not happy about it.

You see, one of physics' most cherished principles is that information can never be truly destroyed. You can scramble it, encode it, or transform it, but like that embarrassing photo you thought you deleted from social media, it always exists somewhere. This principle has worked flawlessly for everything we've ever studied in physics - until we got to black holes.

Black holes seem to break all the rules. Picture a cosmic vacuum cleaner with an appetite for, well, everything. Matter, light, your overdue library books - once anything crosses the event horizon (the point of no return), it apparently vanishes forever. According to Einstein's equations, everything that falls in gets crushed into a single point of infinite density called a singularity. Poof! Gone. No return address, no forwarding number.

This creates what physicists call the black hole information paradox, and it's been giving them headaches for decades. Stephen Hawking made things even more complicated when he showed that black holes slowly evaporate over time through what we now call Hawking radiation. But this radiation appears completely random - it contains no information about what fell in. It's as if our cosmic librarian isn't just losing books; they're replacing them with random pages of gibberish.

This is where SMUT comes in with a radical proposal: what if black holes aren't cosmic paper shredders at all, but rather nature's most sophisticated hard drives?

Remember those spinor fields we talked about in the last chapter - our universal messengers that need two full rotations to return to their starting point? SMUT suggests that when matter falls into a black hole, its information doesn't vanish into a singularity. Instead, it gets encoded into these spinor fields, preserved in the very fabric of spacetime through its twisted geometry (that's where our friend torsion comes back into play).

Think of it like this: imagine you're watching someone drop a book into a pool of water. From above, the book seems to vanish, and all you can see are ripples on the surface. But the information about the book - its shape, mass, the words on its pages - is actually encoded in those ripples, if you know how to read them. SMUT suggests something similar happens with black holes, but instead of ripples in water, we're talking about patterns in spinor fields.

This is more than just a clever bookkeeping trick. It's a complete reimagining of what black holes are. Instead of being cosmic dead ends where information goes to die, they become nature's ultimate storage devices. They're like the cloud storage of the universe, except instead of storing your vacation photos, they're preserving the fundamental information about everything that's ever fallen into them.

But wait, it gets better. Remember how spinors can carry information across different scales? This means that the information stored in black holes isn't just locked away forever. It could potentially be accessed through subtle patterns in the space around the black hole. We might even be able to detect these patterns in the form of gravitational wave "echoes" when black holes collide.

This new perspective changes everything about how we think about black holes. They're not destroyers of information but preservers of it. They're not cosmic trash compactors but nature's most sophisticated filing system. It's as if we've been looking at libraries and seeing only book-eating monsters, when actually, we're dealing with extremely zealous archivists.

The implications go far beyond black holes themselves. If SMUT is right, it means the universe has built-in safeguards against information loss. It suggests a cosmos that's far more coherent and interconnected than we imagined, where even the most extreme objects play by the universal rules of information preservation.

Think about that the next time you hear about something falling into a black hole. It's not being destroyed - it's being archived in the universe's most secure storage system, written in the language of spinors on pages made of twisted spacetime. Our cosmic librarian can rest easy after all.

In the next chapter, we'll explore another mind-bending aspect of SMUT: how this same framework might eliminate the need for dark matter entirely. But for now, take a moment to appreciate how elegant this solution is. Black holes transform from being physics' biggest troublemakers to being perfect examples of how the universe maintains its continuity. Not bad for objects that were once thought to be nothing but cosmic dead ends.

CHAPTER 4:

Hold My Beer, Said the Universe

"In the beginning, there was a bang," said traditional cosmology.

"Hold my beer," said the universe, and started spinning instead.

If you've been following our journey so far, you've seen how SMUT explains black holes not as cosmic paper shredders but as nature's most obsessive data hoarders. You've met spinors, the universe's secret messengers that need two full rotations to return to their starting point, and you've learned about torsion, which gives spacetime its twist. Now it's time to see how all these pieces fit together to solve one of the biggest mysteries in physics: dark matter.

Or rather, the mystery of why we thought we needed dark matter in the first place.

Here's the problem: when astronomers look at galaxies, they spin way too fast. According to the visible matter we can see - stars, gas, dust, forgotten space keys - these cosmic pinwheels should be flying apart like a badly assembled salad spinner. To explain why they don't, physicists invented dark matter - invisible stuff that provides extra gravity to hold everything together. It's like saying the reason your sock disappeared in the dryer is because of invisible sock-eating gremlins. It might explain the observation, but it feels a bit... desperate.

Enter SMUT, with what might be the most elegant "hold my beer" moment in physics history. Instead of inventing invisible matter, what if we're just seeing the effects of preserved angular momentum from the very beginning of the universe? And what if that beginning wasn't a bang at all, but a spin?

Remember those spinors we talked about, the ones that need two full rotations to return home? What if the universe itself started with such a rotation? Instead of an explosion (the Big Bang), picture a cosmic-scale version of what happens when you're lying in bed, realize you're late for work, and do that weird twist-flip move to get up quickly. We call this the Big Spin.

This isn't just a random replacement. The Big Spin explains why everything in the universe - from electrons to galaxies - has angular momentum (spin). It explains why galaxies form into spiral shapes, why they cluster in spiral-like filaments on the largest scales, and why they rotate the way they do - without needing to invent invisible matter to make the math work.

But here's where it gets really interesting: this same spinning start to the universe also explains why we see more matter than antimatter (it got separated in the spin, like cream rising to the top of milk), why the universe appears to be so uniform on large scales (the spin distributed everything evenly), and even provides a natural mechanism for cosmic inflation (that brief moment of incredible expansion early in the universe's history).

It's all connected. The same principles that preserve information in black holes - spinors and torsion - also explain how angular momentum gets preserved across cosmic scales. The same geometric twists that allow for smooth transitions between quantum and classical realms also explain why galaxies rotate the way they do. It's like discovering that your sock didn't disappear because of dryer gremlins, but because of the same principles that govern how your washing machine spins.

This is what makes SMUT so compelling. It's not just a collection of clever solutions to individual problems. It's a framework that shows how all these phenomena are deeply interconnected. The universe isn't a bunch of separate puzzles needing separate solutions - it's one grand, spinning, twisting dance, where the same basic principles manifest at every scale.

Think about what this means for our understanding of the cosmos. Instead of a universe that needs different explanations for different phenomena - dark matter for galaxy rotation, inflation for cosmic uniformity, separate mechanisms for matter-antimatter asymmetry - we have a unified picture where all these features emerge naturally from the same underlying principles.

It's as if we've been looking at a vast tapestry through a series of tiny keyholes, inventing separate explanations for each patch we could see. SMUT suggests that if we step back, we'll see that all these apparently separate phenomena are actually part of one beautiful, coherent pattern.

And it all started with a spin rather than a bang. Sometimes the universe's "hold my beer" moments turn out to be its most profound revelations.

CHAPTER 5:

The Universe's Evil Twin

Every superhero story needs a supervillain. Every Luke Skywalker needs a Darth Vader. And according to SMUT, our universe needs its antimatter twin. That's right - our entire universe might have a twin, and like any good evil twin, it's made of exactly the opposite stuff we are.

Remember how in the last chapter we learned that the universe began not with a bang but with a spin? Well, it turns out that spin did more than just get the cosmic party started. Like a cosmic centrifuge, it separated matter and antimatter into two separate universes. Think of it as the ultimate family estrangement story - one universe got all the matter, the other got all the antimatter, and they've been avoiding each other at cosmic family reunions ever since.

This isn't just a wild idea dreamed up for dramatic effect. It actually solves one of the biggest mysteries in physics: why is our universe made almost entirely of matter? Where did all the antimatter go? Physicists have been scratching their heads over this for decades. According to our best theories, the Big Bang should have created equal amounts of matter and antimatter, which should have immediately annihilated each other in a cosmic fireworks show, leaving nothing but energy behind. Yet here we are, made of matter, living in a matter universe, posting matter selfies on matter social media.

SMUT suggests a beautifully simple solution: the antimatter didn't disappear - it's all hanging out in our twin universe, posting its own antimatter selfies on its antimatter Instagram. The Big Spin separated matter and antimatter as naturally as a salad spinner separates water from lettuce. (Though I wouldn't recommend trying to separate matter and antimatter in your kitchen - the results would be considerably more explosive than damp lettuce.)

But here's where it gets really interesting. Remember those spinor fields we talked about earlier - our universal messengers that need two full rotations to return to their starting point? That 720-degree rotation isn't just a mathematical curiosity. It's exactly what you need to describe a complete matter-antimatter cycle. The first 360 degrees gives you our universe, and the second 360 degrees gives you the antimatter twin.

This isn't just elegant mathematics - it has real implications. For one thing, it means that the total amount of "stuff" in both universes combined perfectly balances out. All the quantum numbers, all the conservation laws that physicists hold dear, they all work perfectly when you consider both universes together. It's like discovering that your checking account isn't actually empty - half your money is just in a parallel account you didn't know about. (Note: Your bank will not accept this explanation for overdraft fees.)

But wait, there's more! This twin universe model also explains why we can't create or find magnetic monopoles (single magnetic poles without their opposite partner). It's not because they don't exist - it's because they're like cosmic joint custody kids, with one pole in our universe and the other in our twin. The mathematics of spinors naturally prevents them from existing entirely in one

17

universe or the other.

Now, you might be wondering: if we have an antimatter twin universe out there, why can't we see it? Well, that's the genius of the separation. The two universes are connected only through gravity and possibly through very subtle quantum effects. It's like having a neighbor who you never see because they work the exact opposite schedule as you - except in this case, they're made of antimatter and live in a parallel universe. (Still probably better than some actual neighbors, though.)

Could we ever detect this twin universe? SMUT suggests we might. Those spinor fields that connect the two universes might leave subtle signatures in gravitational waves, creating echo patterns that our detectors could potentially pick up. It would be like overhearing your antimatter twin's cosmic phone conversation through the walls of reality.

The implications are mind-boggling. Every galaxy in our universe might have an antimatter counterpart in the twin universe. Every planet, every star, possibly even every one of us might have an antimatter doppelganger going about their business in the mirror universe. (Though if sci-fi has taught us anything, it's probably best not to meet your antimatter twin - they usually turn out to be evil and have a goatee.)

This isn't just a solution to the matter-antimatter asymmetry problem - it's a complete reimagining of the cosmic family tree. Our universe isn't an only child; it's one half of a twin pair, separated at birth by the Big Spin, each going their own way but forever connected by the subtle threads of spinor fields and gravity.

And somewhere out there, in that antimatter universe, there might be an antimatter physicist writing a book about their mysterious matter twin, wondering if we're the evil ones. (Spoiler alert: given humanity's history, they might have a point.)

Next up, we'll explore how this twin universe model connects to the way time flows in both universes. Because if you thought having an evil twin was complicated, wait until you hear about how time itself might have a twin...

CHAPTER 6:

The Universe's Layer Cake

Picture the most ambitious baker in history attempting to create the ultimate layer cake. But this isn't just any cake - it's a cake where each layer follows its own laws of physics, where the filling between the layers acts as a universal translator, and where somehow, impossibly, each layer is both separate from and intrinsically connected to all the others. Welcome to the preservation hierarchy, the universe's own recipe for maintaining order across all scales.

Our cosmic cake has three essential layers, each with its own distinctive flavor of physics:

The Classical Layer (Bottom Layer): This is the foundation, the layer we're most familiar with. It's where Newton's laws work beautifully, where objects follow predictable paths, and where cause clearly leads to effect. This layer preserves all the things physicists love to conserve - energy, momentum, angular momentum. It's like the responsible older sibling of our cosmic family, keeping track of all the accounting.

The Quantum Layer (Middle Layer): Here's where things get weird. In this layer, particles exist in multiple states simultaneously, quantum entanglement allows for "spooky action at a distance," and uncertainty is a fundamental feature, not a bug. This layer preserves quantum information and coherence. Think of it as the artistic teenager of the family, exploring all possibilities simultaneously while somehow maintaining its own strange kind of order.

The Conformal Layer (Top Layer): This is perhaps the most mysterious layer, where the geometry of spacetime itself is preserved. It maintains the smooth structure of the universe across all scales, ensuring that the fabric of reality doesn't develop any rips or tears. It's like the family architect, making sure the whole cosmic house doesn't collapse.

But here's the really clever part: these layers aren't just stacked on top of each other like a regular cake. They're more like a quantum cake where each layer exists simultaneously in multiple places, interacting with the others through what we might call "cosmic frosting" - our old friends the spinor fields.

Remember those spinor messengers that need two full rotations to return to their starting point? They're like the secret ingredient that allows these layers to communicate and work together. They carry information between layers, ensuring that what's preserved in one layer is translated appropriately to the others.

Think of it this way: imagine you're watching a ballet performance. The Classical Layer is like seeing the graceful movements of the dancers from your seat in the audience. The Quantum Layer is like simultaneously seeing all the possible positions each dancer could take at any moment. And the Conformal Layer is like the underlying choreography that ensures all these movements work together to create a coherent performance.

The preservation hierarchy isn't just a convenient organizational chart - it's the fundamental structure that allows the universe to maintain continuity across vastly different scales. It's what enables a black hole to preserve information (through all three layers working together), what allows quantum effects to influence cosmic structures (through inter-layer communication), and what makes the Big Spin possible (by coordinating preservation across all layers).

When matter falls into a black hole, for example, its classical properties (mass, charge, angular momentum) are preserved in the Classical Layer. Its quantum information is preserved in the Quantum Layer. And the smooth structure of spacetime around the black hole is preserved in the Conformal Layer. All three layers work together, connected by spinor fields, to ensure that nothing is truly lost.

This layered structure also explains why our everyday experience seems so different from the quantum realm. It's not that quantum mechanics stops working at large scales - it's that the Classical Layer becomes more dominant in our everyday experience, while the Quantum Layer's effects become more subtle (but never disappear completely).

The preservation hierarchy is like the universe's own version of checks and balances. Each layer has its own domain and its own type of preservation, but none can operate entirely independently of the others. They're bound together by spinor fields and torsion, creating a cosmic democracy where every scale of the universe gets a vote.

This might sound complex - and it is! But it's a complexity born of elegance. Just as a master chef knows that the perfect cake requires each layer to complement the others, the universe has found a way to make different types of physical preservation work together harmoniously.

And here's the mind-bending part: this hierarchy might not just preserve physical quantities - it might also preserve the very possibility of existence itself. The Classical Layer preserves the reality we can directly observe, the Quantum Layer preserves the realm of possibility, and the Conformal Layer preserves the stage upon which it all plays out.

Next time someone asks you about the fundamental structure of reality, you can tell them it's like a cosmic layer cake - but one where each layer is infinite, where the frosting carries messages between layers, and where the whole thing is both carefully preserved and constantly evolving.

Just don't try to eat it. Cosmic layer cake is strictly for theoretical consumption only.

CHAPTER 7:

Time Isn't What It Used To Be

"Time is nature's way of keeping everything from happening at once," goes an old physics joke. But according to SMUT, time might be more like nature's way of making everything happen in exactly the right way across different scales - even if that means happening "at once" in some layers and "one thing after another" in others.

Picture time as a river. That's how we usually think of it - flowing smoothly from past to future, carrying everything along in its current. That's a nice, comfortable image. It's also about as accurate as describing the Amazon River as "a bit of water." Just as our cosmic layer cake showed us how space itself has different layers with different properties, SMUT suggests that time itself comes in layers too.

Let's break it down by visiting each layer of our preservation hierarchy to see how time behaves:

In the Classical Layer (our everyday world), time behaves like that river we imagined - flowing steadily in one direction, past to future, cause to effect. This is the time we're familiar with, where you can't unscramble an egg or unsend that embarrassing email to your entire company. It's why we remember the past but not the future, why movies play forward but not backward (except for that one Christopher Nolan film - you know the one).

But climb up to the Quantum Layer, and time gets... peculiar. Here, time becomes more like a lake than a river. Quantum processes can be reversible - run them forward or backward, and they work exactly the same way. It's like having a video that makes perfect sense whether you play it forward or backward. This is why quantum mechanics can seem so weird - it's operating with a different kind of time than we're used to.

And then we reach the Conformal Layer, where time becomes so flexible it's barely recognizable as time anymore. Here, time is more like a cloud than a river or lake - it can reshape itself, connect distant points, and maintain the geometric structure of spacetime itself. If the Classical Layer's time is a straight line and the Quantum Layer's time is a two-way street, the Conformal Layer's time is more like a pretzel - twisted, connected back to itself, but somehow maintaining a perfect geometric form.

This might sound like pure mathematical abstraction, but it has profound implications for how we understand reality. Think about causality - our deep-seated belief that causes must precede effects. In the Classical Layer, this is absolutely true. You can't answer a phone call before it rings (no matter how many times we've all wished we could).

But in the Quantum Layer, cause and effect can get fuzzy. Quantum particles can become "entangled," seemingly influencing each other instantaneously across vast distances. It's not that they're breaking the speed limit of the universe (nothing can travel faster than light) - it's that in the Quantum Layer, time itself works differently.

And in the Conformal Layer? Causality becomes even stranger. Here, it's not about what happens before or after - it's about what preserves the geometric structure of spacetime itself. It's like asking what comes first in a circle - the question itself stops making sense.

This layered view of time helps explain some of the deepest mysteries in physics. Why does time seem to flow in only one direction in our everyday experience, when most fundamental physics equations work equally well forward and backward? Because we're experiencing the Classical Layer's version of time. Why do quantum experiments give such weird results? Because they're operating in a layer where time is reversible.

But perhaps the most mind-bending implication is what this means for free will and the nature of reality itself. If time behaves differently in different layers, but all layers are operating simultaneously and interconnected through spinor fields, what does that mean for cause and effect? For our ability to make choices? For the very nature of what we call "now"?

SMUT suggests that what we experience as the flow of time is actually the interplay between these different temporal layers. The Classical Layer gives us our familiar one-way flow, the Quantum Layer allows for the possibility of multiple futures existing simultaneously, and the Conformal Layer ensures that whatever happens preserves the fundamental structure of spacetime.

It's a bit like being in a jazz band. The rhythm section (Classical Layer) keeps steady time, the soloists (Quantum Layer) can play around with timing and explore different possibilities, and the overall musical structure (Conformal Layer) ensures it all holds together as a coherent piece.

This doesn't mean we can build time machines or change the past (sorry, aspiring time travelers). What it does suggest is that time is far richer and more complex than we imagined. It's not just a river flowing from past to future - it's a symphony of different temporal behaviors working together across different scales of reality.

Next time someone asks you what time it is, you can say "Which layer?" Just don't blame us if that makes you late for dinner.

CHAPTER 8:

CSI: Cosmic Science Investigation

"Pics or it didn't happen!"

If you've spent any time on the internet, you know this classic challenge. Well, the universe doesn't respond to internet memes (usually), but the principle is the same. We've spent several chapters telling you about spinors doing cosmic gymnastics, time existing in layers like some interdimensional cake, and our universe having an antimatter evil twin. Now it's time for the universe to show us the receipts.

Welcome to CSI: Cosmic Science Investigation, where we're hunting for the universe's fingerprints, footprints, and occasionally its selfies. Because SMUT isn't just a pretty theory - it's making some bold claims about what we should actually see out there in the cosmic wild.

Let's start with perhaps the coolest prediction: gravitational wave echoes. You know how when you yell into a canyon, you hear your voice bounce back? SMUT says something similar happens when black holes collide, except instead of sound waves bouncing off canyon walls, we're talking about gravitational waves bouncing through different layers of reality itself.

LIGO (our gravitational wave detector) has already heard black holes go "CHIRP!" when they collide. But SMUT says we should also hear "chirp... chirp... chirp..." - cosmic echoes as these waves reverberate through the classical, quantum, and conformal layers. It's like the universe's version of surround sound!

But wait, there's more! Remember our friend the Big Spin? (Much cooler than the Big Bang - it's like the universe started with a dance move instead of an explosion.) If SMUT is right, this cosmic pirouette should have left traces in the cosmic microwave background radiation - the universe's baby picture. We're talking about subtle patterns that say "I started with a spin" rather than "I started with a bang." It's like finding the universe's baby footprints, but the footprints are doing a twirl.

And galaxies? Oh boy. You know how we currently think galaxies need dark matter to explain why they spin the way they do? SMUT says, "Hold my cosmic beverage - we can explain that with preserved angular momentum!" It's suggesting that galaxies spin the way they do because they're following patterns set up by that initial Big Spin, passed down through the preservation hierarchy like some sort of cosmic hereditary dance moves.

Here's where it gets really wild: these patterns should show up everywhere, at every scale, like the universe is obsessed with a particular screensaver pattern. From electrons to planets to galaxies, we should see similar rotation patterns - a cosmic game of "Simon Says" where everything follows the same basic moves, just at different scales.

And let's not forget about our antimatter evil twin universe!

While we can't exactly peek over the cosmic fence to wave at our antimatter neighbors, SMUT says there should be subtle signs of their gravitational influence. It's like trying to detect your neighbor's activity by feeling tiny vibrations through your apartment wall - if your apartment was the size of a universe and your neighbor was made of antimatter.

So what exactly are we looking for? Here's our cosmic scavenger hunt list:

1. Gravitational Wave Greatest Hits:
    a. Initial "chirp" followed by a series of encores
    b. Each echo spaced out like a cosmic rhythm section
    c. The universe's version of a three-layer acoustic reflection
2. The Universe's Baby Dance Photos:
    a. Subtle swirl patterns in the cosmic microwave background
    b. Evidence of that first cosmic spin move
    c. The original universal dance floor
3. Galaxy Got Moves:
    a. Rotation patterns that don't need dark matter to make sense
    b. Similar spins at different scales
    c. Everything following the same cosmic choreography
4. Quantum-Scale Dance Lessons:
    a. Particles doing 720-degree pirouettes
    b. Quantum systems showing off their preservation moves
    c. The smallest dancers following the biggest patterns
5. Signs of Our Evil Twin:
    a. Gravitational whispers from the antimatter side
    b. High-energy particles playing interdimensional ping-pong
    c. Quantum entanglement reaching across the universal divide

Some of these we can look for right now. Others might need us to build more sensitive instruments - like trying to hear a pin drop in a rock concert, but the pin is quantum-sized and the concert is the entire universe.

The beautiful thing about these predictions is that they're specific enough to be wrong. That's actually crucial in science - a theory needs to stick its neck out and say "if I'm right, you should see THIS." SMUT is practically jumping up and down, pointing at specific places in the cosmos and saying "Look here! And here! And especially here!"

If we find these patterns exactly where SMUT predicts? That's like the universe liking and sharing our theory. If we don't find them, or find something completely different? Well, that's science too. As the great Richard Feynman once said, "If it disagrees with experiment, it's wrong!" (Though he probably said it with more bongo drums in the background.)

We're cosmic detectives on the ultimate scavenger hunt, looking for the universe's autograph written in gravitational waves, its dance moves preserved in galaxy rotations, and maybe even a wave hello from our antimatter twins. The evidence is out there - we just need to know where to look and build instruments sensitive enough to see it.

Just remember: whatever we find, it's bound to be amazing. Because even if SMUT turns out to be wrong, the fact that we can even ask these questions, build these instruments, and go looking for these cosmic signatures is pretty incredible. We're not just passive observers of the cosmic dance - we're learning to read the choreography!

Next up: we'll put all these pieces together and ponder what it means if SMUT is right. Spoiler alert: it's going to blow your mind. Again.

CHAPTER 9:

Stop Looking for Your Keys Under the Lamppost

There's an old joke about a man searching for his lost keys under a streetlight. A passerby asks if he's sure he dropped them there. "No," the man replies, "I lost them in the dark alley over there, but the light's better here."

For decades now, theoretical physics has been that man, searching where it's comfortable rather than where the answers might actually be. We've invented entire categories of undetectable matter, added numerous dimensions we can never observe, and created increasingly baroque mathematical structures to patch the holes in our understanding - all because we're afraid to step away from the lamppost of conventional thinking.

String theorists have spent over 40 years developing beautiful mathematics that, so far, have produced exactly zero testable predictions. Dark matter researchers have built ever-more-sensitive detectors to find particles that stubbornly refuse to exist. We keep adding epicycles to our theories instead of considering that maybe, just maybe, we're thinking about the problem all wrong.

SMUT isn't asking you to believe in invisible dimensions or undetectable particles. It's asking you to look at what we already know and understand it in a new way:

- We know angular momentum is conserved
- We know spacetime can curve
- We know quantum mechanics works
- We know spinors exist
- We know information should be preserved

Instead of inventing new things to explain what we observe, SMUT suggests we need to understand the deeper connections between what we already know is true. The universe isn't hiding its secrets from us - we've just been too busy searching under the lamppost to see them.

Look at the evidence:

- Galaxies spin too fast? Instead of inventing dark matter, consider preserved angular momentum from the Big Spin
- Information vanishing into black holes? Instead of breaking physics, look at how spinor fields preserve it
- Quantum mechanics and gravity won't play nice? Instead of adding dimensions, understand how the preservation hierarchy connects them

The universe is telling us something profound: everything is connected, everything is preserved, and everything follows patterns that repeat across scales. We don't need more complexity - we need a deeper understanding of the simplicity that's already there.

To our colleagues still searching under the lamppost: We understand. The light is better there. The math is familiar there. The funding is there. But the answers aren't there. They're out in the darkness where we need to think differently, where we need to challenge our assumptions about how the universe works.

SMUT isn't just another theory - it's a call to action. Stop inventing things that don't exist. Start understanding the things that do. The universe doesn't need more dimensions or invisible particles. It needs us to open our eyes and see the patterns it's been showing us all along.

The choice is yours: Keep searching under the lamppost, adding epicycles to epicycles, hoping that the next particle detector or the next mathematical framework will finally reveal what's missing...

Or step into the darkness with us. Look at the universe with fresh eyes. See how everything connects, how information persists, how patterns repeat across scales. The answers are there - they always have been. We just needed to think differently to see them.

The future of physics isn't in adding more complexity. It's in understanding the profound simplicity that's been staring us in the face. SMUT is one step in that direction. Who's ready to take the next?

After all, as the saying goes: The definition of insanity is doing the same thing over and over and expecting different results. Isn't it time we tried something new?

The universe is waiting. The patterns are there. The evidence is mounting.

The only question is: Are you ready to think differently?

THE NEW UNIFIED THEORY OF SPINOR-MEDIATED UNIVERSAL TRANSITIONS

Justin Sirótin

Auto-didactic Theoretical Physicist
Independent Researcher

"If I have seen further, it is by standing on the shoulders of giants"
— Sir Isaac Newton

"We cannot solve our problems with the same thinking we used when we created them."
— Albert Einstein

"The most profound discoveries often start with three words: 'This isn't working'"
—Justin Sirotin

ABSTRACT

In the quest to unify quantum mechanics and general relativity, we introduce the Spinor-Mediated Universal Transition (SMUT) model, a transformative framework that bridges cosmic and quantum scales. Building upon the foundational work of Newton and Einstein, this model synthesizes their insights to reveal a self-sustaining, circular structure in the universe that echoes Newton's equal and opposite reactions while incorporating Einstein's spacetime geometry. The SMUT model leverages **spinor mathematics** and **torsion dynamics** to mediate transitions within spacetime at critical densities, sidestepping the singularities that challenge traditional physics. Through spinor fields—unique in their 720-degree rotational symmetry—the model suggests an intrinsic bridge between classical and quantum domains, allowing for a dynamic that preserves continuity across all scales of the universe.

Central to this framework is the **Intermediated Universal Transition**, which introduces a layered **preservation hierarchy** consisting of three interacting layers:

1. **Classical Layer** – Conserves fundamental quantities such as angular momentum and energy.
2. **Quantum Layer** – Preserves coherence and quantum information, maintaining consistency without loss across transitions.
3. **Conformal Layer** – Ensures geometric continuity, creating smooth transitions that avoid the breakdowns predicted by classical theories.

Each layer activates sequentially, harmonizing the conservation laws and coherence principles that guide cosmic evolution.

This hierarchical coupling between spinors and spacetime torsion provides a natural solution to longstanding cosmological challenges, such as black hole singularities, dark matter, matter-antimatter asymmetry, and the origin of the universe. Through this spinor-torsion mechanism, the model enables the universe to self-regulate, revealing a regenerative, cyclic structure where boundaries of one scale flow seamlessly into the next.

The SMUT model also makes clear and testable predictions across multiple observational domains. **Gravitational wave signatures** emerge as distinctive resonance frequencies and echo cascades, potentially detectable in current observatories. Long-term studies of **black hole spin distributions** may reveal evolutionary patterns that deviate from traditional models, consistent with spinor-torsion dynamics. The model aligns with Big Bang Nucleosynthesis predictions, offering solutions to the **lithium-7 discrepancy** and supporting observed baryon asymmetry through CP-violating spinor interactions. Finally, this framework provides a potential reinterpretation of dark matter as a geometric effect of spinor-preserved angular momentum rather than an unknown form of matter.

With its ability to link quantum interactions with cosmic-scale structures, the SMUT model presents a unified, observationally testable theory that redefines our understanding of continuity, structure, and conservation across the universe's vast scales. This theory not only unites classical and quantum realms but suggests a deeper, interconnected order underlying the cosmos, inviting a new era of experimental validation to explore the nature of reality itself.

I. INTRODUCTION

A. The Legacy of Newton and Einstein

Our understanding of the universe's fundamental structure has long rested on two towering frameworks: Newton's laws of motion and Einstein's general relativity. Newton's work established a model of nature governed by predictable interactions between bodies, a profound insight that shaped physics for centuries. His **third law of motion**—that for every action there is an equal and opposite reaction—provides a core principle of balance and reciprocity. This principle underpins the interactions we observe on a daily scale, from the mechanics of celestial bodies to the forces at play within simple machines, setting a foundation that still holds in many realms of modern science.

Einstein's work on **general relativity** expanded our view, shifting focus from forces acting at a distance to the curvature of spacetime itself. In Einstein's vision, gravity was no longer a simple force but a manifestation of spacetime curvature caused by the presence of mass and energy. This framework allowed for predictions of phenomena previously unimagined—gravitational waves, the bending of light near massive objects, and the expansion of the universe—all validated through observational evidence. Einstein's insight redefined our perception of reality, revealing that spacetime is dynamic and responsive, curving and stretching in the presence of matter.

However, as profound as these frameworks are, neither fully encompasses the fundamental nature of reality, especially at the extremes of scale. Newton's laws falter when describing the subatomic realm, where quantum effects dominate and particles behave probabilistically rather than deterministically. Similarly, general relativity, while comprehensive in describing large-scale cosmic phenomena, encounters paradoxes when applied to singularities within black holes and the conditions of the early universe.

These unresolved contradictions highlight the need for a new theory—one that can seamlessly integrate **quantum mechanics** and **general relativity** into a unified framework. Such a theory must be able to describe the interactions of both particles and planets, reconciling the inherent unpredictability of quantum mechanics with the geometric elegance of spacetime curvature. In this pursuit, we build upon the legacies of Newton and Einstein, aiming to transcend their limits and create a theory that harmonizes the forces and phenomena across all scales, from the smallest particles to the vast stretches of the cosmos.

B. The Spinor-Mediated Transition Mechanism

The **Spinor-Mediated Transition Mechanism** extends Newtonian and Einsteinian principles into a unified framework, creating a **circular, regenerative structure** that bridges the classical and quantum realms. By integrating **spinor fields** with **torsion geometry** within spacetime, this mechanism enables continuous transitions across scales, addressing longstanding paradoxes and preserving fundamental laws.

Spinor fields, with their 720-degree rotational symmetry, serve as mediators between the quantum and classical domains, carrying information across transitions. These fields, coupled with spacetime torsion, allow for smooth transformations where classical theories would otherwise predict singularities or discontinuities. Torsion introduces an intrinsic "twist" in spacetime, permitting the fabric of the universe to dynamically adapt under extreme conditions, such as those near black holes or during the universe's early moments.

Through a layered **preservation hierarchy**, the model resolves paradoxes across multiple domains:

- **Fundamental Paradoxes**: Preserves information in black holes, addresses matter-antimatter asymmetry through paired universe production, and reframes quantum measurement as a preservation transition.
- **Cosmological Paradoxes**: Solves the horizon and flatness problems by maintaining causal connections and natural geometric structure.
- **Gravitational Paradoxes**: Avoids singularities with smooth transitions at critical radii and interprets dark energy as a natural result of the universe's initial "Big Spin."
- **Quantum Paradoxes**: Maintains non-local connections through the hierarchy, ensuring coherence and preserving CP symmetry across paired universes.
- **Structural Paradoxes**: Explains galaxy formation and dark matter as results of preserved angular momentum patterns.
- **Theoretical Paradoxes**: Provides a natural basis for the arrow of time, quantum-classical transition, and the origin of physical constants.

With these mechanisms, the Spinor-Mediated Transition Mechanism not only unifies forces across scales but also predicts observable phenomena, such as gravitational wave echoes, specific cosmic structure patterns, and CMB signatures, laying the groundwork for experimental validation of this seamless, interconnected model.

C. Objectives and Scope of the Paper

This paper aims to introduce and rigorously explore the **Spinor-Mediated Transition Mechanism** as a unifying framework that bridges quantum mechanics and general relativity through a layered **preservation hierarchy**. The primary objectives are to:

1. **Present the Preservation Hierarchy**: The paper will outline the structure and function of the preservation hierarchy's three layers—classical, quantum, and conformal—which work in tandem to maintain continuity, coherence, and conservation across scales. Each layer will be examined in terms of its role in ensuring smooth transitions and preserving fundamental physical principles, from angular momentum to quantum information.

2. **Address Key Cosmological Challenges**: The model's implications for resolving longstanding paradoxes and challenges in physics will be detailed, including:
   a. **Black Hole Information Paradox**: Proposing a mechanism for information preservation through spinor fields and torsion.
   b. **Matter-Antimatter Asymmetry**: Introducing paired universe production to balance matter and antimatter.
   c. **Horizon and Flatness Problems**: Showing how initial "Big Spin" correlations and preserved geometry provide natural resolutions.
   d. **Singularity and Dark Energy Problems**: Demonstrating how smooth transitions and the regenerative structure of the universe maintain continuity without singularities and support cosmic expansion.

3. **Preview Observational Predictions**: To validate the model, the paper will preview a set of testable predictions that offer clear paths for experimental verification:
   a. **Gravitational Wave Echoes**: Detectable echoes and

resonance patterns from the preservation hierarchy's sequential transitions.
   b. **Cosmic Microwave Background (CMB) Signatures**: Distinct patterns and alignments in the CMB arising from the initial Big Spin transition.
   c. **Galaxy Rotation and Structure Patterns**: Observable rotation dynamics and structural patterns reflecting preserved angular momentum across scales.
4. **Establish Testable Outcomes**: Each prediction will be linked to specific observational strategies, laying out potential experimental approaches using gravitational wave observatories, CMB analysis, and large-scale cosmic surveys to test the model's foundational principles.

Through this paper, we aim to provide a coherent and mathematically consistent framework for exploring and validating the Spinor-Mediated Transition Mechanism, moving toward a comprehensive theory that integrates classical and quantum principles across all scales of the cosmos.

II. Mathematical Framework

A. Core Spinor-Torsion Mechanism

The fundamental mechanism of our theory rests on the interaction between spacetime torsion and spinor fields, mediated through a precise mathematical framework that extends Einstein-Cartan theory.

1. Modified Einstein-Cartan Equations

The field equations incorporating spinor-torsion coupling take the form:

$$R_{\mu\nu} - (1/2) R g_{\mu\nu} = 8\pi G (T_{\mu\nu}(matter) + T_{\mu\nu}(spin))$$

where:
$$T_{\mu\nu}(\text{spin}) = \kappa(\bar{\psi}\gamma_{(\mu}\nabla_{\nu)}\psi - \nabla_{(\mu}\bar{\psi}\gamma_{\nu)}\psi)$$

Modification includes:
- Torsion contribution to curvature
- Spinor field energy-momentum
- Cross-coupling terms

2. Spinor Field Coupling

The spinor-torsion interaction is described by:

Modified connection
$$\omega_{\mu ab} = \omega_{\mu ab}(LC) + K_{\mu ab}$$

Contortion tensor
$$K_{\mu\nu\lambda} = -(1/2)(S_{\mu\nu\lambda} + S_{\nu\mu\lambda} - S_{\lambda\mu\nu})$$

Spinor covariant derivative
$$D_\mu\psi = \partial_\mu\psi + (1/4)\omega_{\mu ab}\gamma^a\gamma^b\psi$$

Critical coupling at rc
$$S_{\mu\nu\lambda}|_{r=r_c} = \eta(\epsilon_{\mu\nu\lambda\sigma}J^\sigma)/\rho$$

3. Conservation Laws

The framework maintains fundamental conservation through:

Total angular momentum conservation
$$\nabla_\mu(J^{\mu\nu} + S^{\mu\nu}) = 0$$
where:
$$S^{\mu\nu} = (1/2)\bar{\psi}\gamma^{\mu\nu}\psi$$
$$\gamma^{\mu\nu} = (i/2)[\gamma^\mu,\gamma^\nu]$$

Energy-momentum conservation

$$\nabla\mu T\mu\nu = S\lambda\mu\nu \nabla\lambda(\bar{\psi}\gamma\mu\psi)$$

Spinor current conservation
$$\nabla\mu(\bar{\psi}\gamma\mu\psi) = 0$$

4. Critical Behavior

At the transition point rc, these equations manifest as:

Geometric transition
$$ds^2|r=rc = ds^2 standard[1 + \kappa(\bar{\psi}\gamma\mu\psi)S\mu]$$

Phase evolution
$$\Phi(r) = \Phi 0 \exp(2\pi i\theta/720°) \quad // \text{ Full spinor rotation}$$

Information preservation
$$S(out) = S(in) + \int rc(dS/dr)transition\, dr$$

This core mechanism provides the mathematical foundation for:
1. Smooth transitions at critical points
2. Information preservation
3. Angular momentum conservation
4. Geometric continuity

The beauty of these equations lies in their natural emergence from fundamental principles and their ability to maintain continuity across all scales through the preservation hierarchy.

B. Preservation Hierarchy

The universe maintains continuity through three interconnected layers of preservation, each building upon and supporting the others through precise mathematical relationships.

1. Classical Layer Mathematics

Primary conservation equations

$\nabla\mu(J\mu\nu + S\mu\nu) = 0$ // Angular momentum
$\nabla\mu T\mu\nu = S\lambda\mu\nu \nabla\lambda(\bar{\psi}\gamma\mu\psi)$ // Energy-momentum

Layer activation function
  $f_1(r) = \tanh((r_c - r)/\lambda_1)$
  where $\lambda_1 = l_p\sqrt{(M/m_p)}$ // Classical scale

Modified force law
  $F\mu = F\mu(\text{standard})[1 + \kappa_1(\bar{\psi}\gamma\mu\psi)S\mu]f_1(r)$

2. Quantum Coherence Preservation

Quantum state evolution
  $D(S\_ent)/d\tau = \eta \nabla\mu(S\mu\nu\lambda\bar{\psi}\gamma\nu\psi)$

Coherence maintenance
  $|\psi(t)\rangle = \exp(iHt/\hbar)|\psi_0\rangle[1 + f_2(r)]$
  where $f_2(r) = \tanh((r_c - r)/\lambda_2)$
  $\lambda_2 = l_p(M/m_p)^{(1/3)}$ // Quantum scale

Phase preservation
  $\Phi_{out} = \exp(4\pi i)\Phi_{in}$ // Full 720° rotation

3. Conformal Geometric Continuity

Metric preservation
$[g\mu\nu]_{r=rc} = 0$ // Continuity
$K_{ij}|_{r=rc+} = -K_{ij}|_{r=rc-}$ // Flip

Conformal transformation
$\tilde{g}\mu\nu = \Omega^2(x)g\mu\nu$
where $\Omega(x) = 1 + f_3(r)$
$f_3(r) = \tanh((rc - r)/\lambda_3)$

Structure preservation
$\nabla\mu(c\tilde{R}g\mu\nu + K\mu\nu) = 0$

4. Layer Coupling

Inter-layer resonance
$\omega_{12} = c/\sqrt{(\lambda_1\lambda_2)} = (c/lp)(mp/M)^{\wedge}(1/4)$ // Classical-Quantum
$\omega_{23} = c/\sqrt{(\lambda_2\lambda_3)} = (c/lp)(mp/M)^{\wedge}(1/3)$ // Quantum-Conformal

Combined activation
$F\_total = F_1f_1(r) + F_2f_2(r) + F_3f_3(r)$

Cross-coupling terms
$\kappa_{ij} = \alpha_{ij}(M/mp)^{\wedge}(\beta_{ij}) * \exp(i\omega_{ij}\tau)$

5. Cascade Activation

Sequential timing
$\tau_1 = GM/c^3$ // Classical activation
$\tau_2 = \tau_1(\lambda_2/\lambda_1)$ // Quantum activation
$\tau_3 = \tau_1(\lambda_3/\lambda_1)$ // Conformal activation

> **Information flow**
> $dI/d\tau = \sum_i (dI/d\tau)_i f_i(r)$

This hierarchy ensures:

1. Conservation laws remain valid at all scales
2. Quantum information is preserved through transitions
3. Geometric continuity is maintained
4. Different scales remain causally connected
5. Information flows smoothly between layers

The beauty of this structure lies in how each layer supports and enables the others, creating a robust framework for maintaining continuity across all scales of the universe.

C. Transition Dynamics

The behavior at critical radius rc represents the core mechanism of our framework, where the preservation hierarchy enables smooth transitions across seemingly singular points.

1. Critical Radius Behavior

Critical radius definition
$$r_c = 2GM/c^2[1 + \kappa(\bar{\psi}\gamma\mu\psi)S\mu]$$

Metric behavior at transition
$$ds^2|r=r_c = ds^2_{standard}[1 + f(transition)]$$
where:
$$f(transition) = \kappa(\bar{\psi}\gamma\mu\psi)S\mu + \eta \int S\mu\nu\lambda(J\sigma K\mu\nu)d\Sigma\lambda$$

Phase evolution through rc
$$\Phi(r) = \Phi_0 \exp(2\pi i\theta/720°) * \{$$
$\quad$ r > rc: standard evolution
$\quad$ r = rc: transition phase
$\quad$ r < rc: mirror phase
$\}$

2. Angular Momentum Conservation

Total angular momentum preservation
$$\nabla\mu(J\mu\nu + S\mu\nu) = 0$$

Detailed balance at rc
$$J\mu\nu|r=r_c+ = J\mu\nu|r=r_c- + \int S\mu\nu\lambda(J\sigma K\mu\nu)d\Sigma\lambda$$

Rotational transformation
$$R(\theta) = \exp(i\sigma \cdot \hat{n}\theta/2) \text{ through } 720°$$
$\{$
$\quad$ 0° to 360°: First universe
$\quad$ 360° to 720°: Twin universe
$\}$

3. Information Preservation Mechanisms

```
Entropy evolution
    dS/dτ = (dS/dτ)standard + κ∫Sμνλ(JσKμν)dΣλ

Information content
    I(r) = I₀[1 + Σᵢfᵢ(r)]
    where fᵢ(r) = preservation layer contributions

Quantum state preservation
    |ψ(τ)⟩ = U(τ)|ψ₀⟩
    U(τ) = T[exp(-i∫H(τ')dτ')] * {
        Layer 1: Classical preservation
        Layer 2: Quantum coherence
        Layer 3: Geometric continuity
    }
```

4. Cross-Scale Coupling

```
// Layer interaction at rc
    λcoupling = √(λ₁λ₂λ₃)

Resonance frequencies
    ω(k) = ω₀√(1 + κ|ψ̄γμψ|²/ρc)

Beat patterns
    fbeat = |ω₁₂ - ω₂₃|
```

5. Observable Signatures

```
Gravitational wave pattern
    h(t) = A₀exp(-t/τ₀) + Σ□ A□[
    exp(-t/τ□)cos(ω□t) +
    exp(-(t-Δt_primary)/τ□)cos(ω□(t-Δt_primary)) +
    exp(-(t-Δt_secondary)/τ□)cos(ω□(t-Δt_secondary))
    ]

Echo delays
```

THE ELEGANT SOLUTION

$$\Delta t\_primary = \tau_0 \log(\kappa |\bar{\psi}\gamma\mu\psi|^2/\rho c)$$
$$\Delta t\_secondary = \tau_0(\lambda_1/\lambda_2)$$

The transition dynamics demonstrate how:

1. Singularities are avoided through smooth transitions
2. Information is preserved across apparent horizons
3. Angular momentum drives geometric transformation
4. Observable signatures emerge naturally
5. Twin universes form through complete rotation

This mathematical structure provides specific predictions for:

- Gravitational wave patterns
- Black hole evolution
- Cosmic structure formation
- High-energy phenomena

III. UNIVERSAL STRUCTURE AND FORMATION

A. Big Spin Model

The Big Spin model replaces the traditional Big Bang singularity with a spinor-mediated transition, providing a natural mechanism for cosmic evolution through preserved angular momentum and geometric transformation.

1. Fundamental Transition Structure

Complete transition description
$$\Psi(\text{universe}) = \Psi_1(0° \text{ to } 360°) + \Psi_2(360° \text{ to } 720°)$$

Phase evolution through transition
$$T(\tau) = T_0 \exp(-\tau/\tau_0)[1 + \kappa(\bar{\psi}\gamma\mu\psi)S\mu]\{$$
  Phase I: Spinor alignment
  Phase II: Geometric transition
  Phase III: Preservation cascade
$$\}$$

Scale factor evolution
$$a(t) = a_0(t/t_0)^{(2/3)}[1 + \eta \int S\mu\nu\lambda(J\sigma K\mu\nu)d\Sigma\lambda]$$

2. Initial Conditions

Critical parameters at transition
$$\rho c = \rho_{\text{Planck}}[1 + f(\text{preservation terms})]$$
$$T_c = T_{\text{Planck}}[1 + g(\text{transition effects})]$$

Angular momentum distribution
$$J(r) = J_0(r/r_0)^{(2/3)}[1 + \kappa(\bar{\psi}\gamma\mu\psi)S\mu]$$

Initial field configuration
$$\Phi(0) = \Phi_0 \exp(i\theta)\{$$
$$\theta \in [0, 4\pi] \text{ // Full spinor rotation}$$
$$\nabla_\mu \Phi = S_{\mu\nu\lambda} \nabla^\lambda \Phi \text{ // Torsion coupling}$$
$$\}$$

3. Evolution Equations

Modified Friedmann equations
$$(\dot{a}/a)^2 = (8\pi G/3)\rho[1 + f(\text{preservation})]$$
$$\ddot{a}/a = -(4\pi G/3)(\rho + 3p/c^2)[1 + g(\text{transition})]$$

Temperature evolution
$$T(r,t) = T_0(t)(r/r_0)^{\wedge}(-3/4)[1 + \kappa_1(\lambda_1/r) + \kappa_2(\lambda_2/r)^2]$$

Density evolution
$$\rho(r,t) = \rho_0(t)(r/r_0)^{\wedge}(-3)[1 + \eta \int S_{\mu\nu\lambda}(J^\sigma K^{\mu\nu})d\Sigma^\lambda]$$

4. Preservation Mechanisms

Conservation equations
$$\nabla_\mu T^{\mu\nu} = S^{\lambda\mu\nu} \nabla_\lambda(\bar{\psi}\gamma_\mu \psi) \text{ // Energy-momentum}$$
$$\nabla_\mu(nBu^\mu) = 0 \text{ // Baryon number}$$
$$\nabla_\mu(J^{\mu\nu} + S^{\mu\nu}) = 0 \text{ // Angular momentum}$$

Entropy evolution
$$dS/dt = (dS/dt)_{standard}[1 + f(\text{preservation})]$$

5. Observable Consequences

CMB temperature fluctuations
$$\delta T/T = (\delta T/T)_{standard}[1 + \kappa(\bar{\psi}\gamma_\mu \psi)S^{\mu\nu\lambda} \nabla_\lambda \Phi]$$

Matter distribution
$$\delta\rho/\rho = (\delta\rho/\rho)_{standard}[1 + \eta(M/M^*)^{\wedge}(1/6)]$$

```
Structure formation
    D(a,k) = D₀(a)[1 + f(k,M*,M₂)]
```

This model provides:

1. Smooth transition instead of initial singularity
2. Natural explanation for cosmic rotation patterns
3. Preservation of quantum correlations
4. Mechanism for matter-antimatter separation
5. Framework for structure formation

Key predictions include:

- Specific CMB patterns
- Large-scale rotation signatures
- Preserved quantum correlations
- Twin universe formation

B. Twin Universe Production

The 720-degree spinor rotation naturally produces paired universes with complementary properties, resolving fundamental symmetry puzzles while maintaining conservation laws.

1. Matter-Antimatter Separation

```
Complete transition wavefunction
    Ψ(total) = Ψ₁(0° to 360°) + Ψ₂(360° to 720°)

Matter-antimatter distribution
    ρ(matter)₁ = +ρ₀[1 + κ(ψ̄γμψ)Sμ]    // Universe 1
    ρ(matter)₂ = -ρ₀[1 + κ(ψ̄γμψ)Sμ]    // Universe 2

Baryon asymmetry in each universe
    η₁ = +6×10^(-10) // Our universe
    η₂ = -6×10^(-10) // Twin universe
    η(total) = 0     // Net symmetry preserved
```

2. Balanced Cp Violation

CP transformation across universes
$CP(\Psi_1) = -CP(\Psi_2)$

Strong CP parameter
$\theta(QCD)_1 = +\theta_0[1 + f(preservation)]$
$\theta(QCD)_2 = -\theta_0[1 + f(preservation)]$

Unified CP violation mechanism
$L(CP) = \kappa CP(\bar{\psi}\gamma_5\gamma\mu\psi)(S\lambda\mu\nu/\rho c)\{$
  Phase(0° to 360°): +CP violation
  Phase(360° to 720°): -CP violation
$\}$

3. Cross-Universe Coupling

Interface dynamics
$\Phi(boundary) = \Phi_0 \exp(2\pi i)[1 + \eta \int S\mu\nu\lambda(J\sigma K\mu\nu)d\Sigma\lambda]$

Entanglement preservation
$S(ent) = S_1 + S_2 = constant$

Information transfer
$dI/d\tau = \pm\kappa(\bar{\psi}\gamma\mu\psi)S\mu\{$
  +: Universe 1 → Universe 2
  −: Universe 2 → Universe 1
$\}$

4. Observable Signatures

Gravitational wave patterns
$h_1(f) = h_0(f)[1 + \kappa(\bar{\psi}\gamma\mu\psi)S\mu]\exp(+i\pi)$
$h_2(f) = h_0(f)[1 + \kappa(\bar{\psi}\gamma\mu\psi)S\mu]\exp(-i\pi)$

Interface phenomena

```
E(boundary) = E₀[1 + f(transition)]{
    Gravitational echoes
    Quantum correlations
    Torsion effects
}

Correlation functions
ξ(r) = ξ₀(r)[1 + γ∫Sμνλ(JσKμν)dΣλ]
```

5. Conservation Laws

```
Total quantum numbers
    B(total) = B₁ + B₂ = 0    // Baryon number
    L(total) = L₁ + L₂ = 0    // Lepton number
    Q(total) = Q₁ + Q₂ = 0    // Electric charge

Angular momentum balance
    J₁ + J₂ = J(total) = constant

Energy conservation
    E₁ + E₂ = E(total)
```

This mechanism provides:

1. Natural resolution of matter-antimatter asymmetry
2. Solution to the strong CP problem
3. Maintenance of fundamental symmetries
4. Testable gravitational wave predictions
5. Framework for cosmic balance

Key implications:

- Perfect symmetry at the grand scale
- Observable interface effects
- Preserved quantum correlations
- Testable gravitational signatures

C. Nested Spherical Geometry

The universe exhibits a fractal structure of nested spherical symmetries, maintained by the preservation hierarchy across all scales.

1. Scale Relativity

```
Scale transformation law
   r' = r[1 + κ(ψ̄γμψ)Sμνλ(Jσ/ρ)]

Nested scale relationships
   λ□₊₁ = λ□(M/mp)^(1/3)[1 + f(preservation)]
   where:
   λ₁ = lp√(M/mp)    // Quantum scale
   λ₂ = lp(M/mp)^(1/3) // Conformal scale
   λ₃ = lp(M/mp)^(1/2) // Classical scale

Observer-dependent metric
   ds² = g̃μν(x,λ(x))dxμdxν
   λ(x) = λ₀[1 + η∫Sμνλ(JσKμν)dΣλ]
```

2. Fractal Preservation Patterns

Self-similar structure function
$P(k,n) = P_0(k)[1 + \alpha(k/k_\square)^{(1/3)}]\exp[-\beta(k/k^*)^2]$
where:
$k_\square = 2\pi/\lambda_\square$ // Scale-dependent wavenumber

Correlation hierarchy
$\xi(r,n) = \xi_0(r)[1 + \gamma_\square \int S\mu\nu\lambda(J\sigma K\mu\nu)d\Sigma\lambda]$

Pattern reproduction
$\Phi_{\square+1}(r) = \Phi_\square(r/\lambda_\square)[1 + f(\text{preservation})]$

3. Geometric Stability

Stability condition
$K(\text{sphere}) = 8\pi G\rho[1 + f(\text{preservation})]$

Angular momentum preservation
$L(n+1) = L(n)[1 + \kappa(\bar{\psi}\gamma\mu\psi)S\mu]$

Nested rotational dynamics
$\Omega(r,n) = \Omega_0(r/r_\square)^{(-2)}[1 + \eta \int S\mu\nu\lambda(J\sigma K\mu\nu)d\Sigma\lambda]$

4. Scale-Invariant Properties

Preserved ratios across scales
$\Omega = \rho/\rho c \approx 1$ // Density parameter
$\eta = n_B/n_\gamma \approx 6\times 10^{-10}$ // Baryon-to-photon ratio
$L/M \propto$ constant // Specific angular momentum

Scale-dependent coupling
$g(\lambda) = g_0[1 + \beta(\lambda/\lambda p)^{(\alpha)}]$

Information preservation
$I(n+1) = I(n)[1 + f(\text{transition})]$

5. Observable Consequences

> Structure formation
> $\delta(k,n) = \delta_0(k)[1 + f(k,\lambda_\square)]D(a)$
>
> Rotational patterns
> $v(r,n) = v_0(r/r_\square)^{\wedge}(1/2)[1 + \kappa(\bar{\psi}\gamma\mu\psi)S\mu]$
>
> Gravitational lensing
> $\psi_{lens} = \psi GR[1 + \eta \int S\mu\nu\lambda(J\sigma K\mu\nu)d\Sigma\lambda]$

Key features:

1. Each scale is self-similar yet unique
2. Preservation hierarchy maintains stability
3. Information flows smoothly across scales
4. Rotational patterns repeat fractally
5. Observable signatures at each level

This structure explains:

- Galaxy formation patterns
- Cosmic web structure
- Dark matter effects
- Gravitational lensing
- Scale-dependent phenomena

IV. RESOLUTION OF CLASSICAL PROBLEMS

A. Dark Matter Reinterpretation

What appears as dark matter emerges naturally from preserved angular momentum and torsion effects through the preservation hierarchy.

1. Angular Momentum Preservation

```
Total angular momentum conservation
    ∇μ(Jμν + Sμν) = 0

Modified effective mass distribution
    M_eff(r) = M(r)[1 + κ(ψ̄γμψ)Sμνλ(Jσ/ρ)]

Preserved rotational dynamics
    L(r) = L₀(r/r₀)^(2/3)[1 + η∫Sμνλ(JσKμν)dΣλ]{
        Galaxy scale: spiral patterns
        Cluster scale: bulk rotation
        Cosmic scale: web structure
    }
```

2. Modified Rotation Curves

```
Velocity profile including preservation effects
    v²(r) = v²_Newton + v²_preservation
    where:
    v²_preservation = κ(ψ̄γμψ)Sμνλ(Jσ/ρ)

Scale-dependent modification
    f(r) = f₀[1 + (r/r*)^α]{
        r < r*: Newtonian dominance
        r ≈ r*: Transition region
```

$r > r^*$: Preservation effects
}

Observable rotation curve
  $v\_obs(r) = \sqrt{(GM(r)/r)}[1 + f(r)]$

3. Structure Formation

Modified growth equation
  $\ddot{\delta} + 2H\dot{\delta} = 4\pi G\bar{\rho}\delta[1 + \eta(M/M^*)^{\wedge}(1/6)]$

Characteristic mass scales
  $M^* = M_p(\lambda_1/l_p)^3$ // Primary transition
  $M_2 = M_p(\lambda_2/l_p)^3$ // Secondary transition

Structure function
  $P(k) = P_0(k)[1 + \alpha(k/k_0)^{\wedge}(1/3)]\exp[-\beta(k/k^*)^{\wedge}2]$

4. Observable Predictions

Gravitational lensing
$\psi_{lens} = \psi_{GR}[1 + \eta \int S\mu\nu\lambda(J\sigma K\mu\nu)d\Sigma\lambda]$

Galaxy cluster dynamics
$\sigma v^2 = \sigma v^2\_standard[1 + \kappa(\bar{\psi}\gamma\mu\psi)S\mu]$

Large scale structure
$\xi(r) = \xi_0(r)[1 + \gamma \int S\mu\nu\lambda(J\sigma K\mu\nu)d\Sigma\lambda]$

5. Testing Framework

```
Key observational tests
    1. Rotation curve shapes
    2. Lensing patterns
    3. Cluster dynamics
    4. Structure correlations
    5. Velocity dispersions

Distinguishing features
    - Scale dependence
    - Angular momentum correlations
    - Geometric patterns
```

This reinterpretation:

1. Eliminates need for exotic dark matter
2. Explains observed rotation curves
3. Maintains structure formation
4. Provides testable predictions
5. Emerges naturally from preservation hierarchy

The apparent effects of dark matter are revealed as manifestations of:

- Preserved angular momentum
- Torsion-spinor coupling
- Geometric preservation
- Scale-dependent dynamics

B. Gravitational Effects

The preservation hierarchy naturally modifies gravitational effects through spinor-torsion coupling and preserved angular momentum.

1. Modified Force Law

Complete gravitational force
$$F = F_{Newton}[1 + \kappa(\bar{\psi}\gamma\mu\psi)S\mu\nu\lambda(J\sigma/\rho)]$$

Scale-dependent coupling
$$G\_eff(r) = G_0[1 + f(r,\lambda)]\{$$
  λ_1: Quantum scale effects
  λ_2: Intermediate coupling
  λ_3: Large-scale behavior
}

Potential modification
$$\Phi(r) = -GM/r + \Phi_{preservation}$$
where:
$$\Phi_{preservation} = \int S\mu\nu\lambda(J\sigma K\mu\nu)d\Sigma\lambda$$

2. Lensing Predictions

Modified deflection angle
$$\alpha = \alpha\_GR[1 + \eta \int S\mu\nu\lambda(J\sigma K\mu\nu)d\Sigma\lambda]$$

Lensing potential
$$\psi_{lens} = \psi_{standard}[1 + f(preservation)]\{$$
  Strong lensing: $\alpha > 1"$
  Weak lensing: $\alpha < 1"$
  Microlensing: stellar scale
}

Magnification matrix
$$\mu_{ij} = \delta_{ij} + \partial_i\partial_j\psi_{lens}[1 + \kappa(\bar{\psi}\gamma\mu\psi)S\mu]$$

3. Cluster Dynamics

Virial theorem modification
$$2\langle T \rangle = -\langle V \rangle[1 + \eta(M/M^*)^{(1/6)}]$$

Velocity dispersion
$$\sigma v^2 = \sigma v^2\_standard[1 + f(preservation)]\{$$
  Core region: $r < r_c$
  Intermediate: $r_c < r < r_{vir}$
  Outer region: $r > r_{vir}$

Mass profile
$M(r) = M\_standard(r)[1 + \kappa \int S\mu\nu\lambda(J\sigma K\mu\nu)d\Sigma\lambda]$

4. Observable Signatures

Distinctive features
1. Scale-dependent effects:
$dG/dr \propto \sum_i f_i(r/\lambda_i)$

2. Angular momentum coupling:
$L(r) = L_0(r/r_0)^{(2/3)}[1 + \text{preservation terms}]$

3. Geometric patterns:
$\xi(r) = \xi_0(r)[1 + \gamma \int S\mu\nu\lambda(J\sigma K\mu\nu)d\Sigma\lambda]$

4. Resonance phenomena:
$\omega(k) = \omega_0\sqrt{(1 + \kappa|\bar{\psi}\gamma\mu\psi|^2/\rho c)}$

5. Testing Framework

Key observations
a) Gravitational wave patterns
$h(f) = h_0(f)[1 + \text{modification terms}]$

b) Lensing surveys
$\kappa(\theta) = \kappa_0(\theta)[1 + \text{preservation effects}]$

c) Cluster catalogs
$M(r)$ vs. $\sigma v(r)$ relationships

d) Large-scale structure
$P(k) = P_0(k)[1 + \text{scale-dependent terms}]$

These modifications:

1. Maintain Einstein's principles
2. Preserve causality
3. Respect conservation laws

4. Provide testable predictions
5. Emerge naturally from framework

Key predictions include:

- Distinctive lensing patterns
- Modified cluster profiles
- Scale-dependent effects
- Observable resonances
- Specific geometric signatures

C. Early Universe Physics

The preservation hierarchy naturally explains early universe phenomena through spinor-mediated transitions and balanced symmetries.

1. Nucleosynthesis Constraints

```
Modified temperature evolution
    T(t) = T₀(t/t₀)^(-3/4)[1 + κ(ψ̄γμψ)Sμ]{
        Phase I:   Spinor alignment
        Phase II:  Element formation
        Phase III: Preservation cascade
    }

Nuclear reaction rates
    Γ(p + n → D + γ) = Γ₀[1 + η∫Sμνλ(JσKμν)dΣλ]

Element abundances
    Y(t) = Y₀exp(-t/τ)[1 + f(preservation)]{
        D/H = (2.547 ± 0.025) × 10⁻⁵
        ³He/H = (1.1 ± 0.2) × 10⁻⁵
        ⁴He: Yp = 0.2472 ± 0.0012
        ⁷Li/H = (1.6 ± 0.3) × 10⁻¹⁰
    }
```

2. Matter-Antimatter Asymmetry

```
Baryon asymmetry through transition
    η = (nB - nB̄)/nγ = 6×10⁻¹⁰[1 + κ(ψ̄γμψ)Sμ]

CP violation mechanism
    L_CP = κCP(ψ̄γ5γμψ)(Sλμν/ρc){
        0° to 360°:   Matter universe
        360° to 720°: Antimatter universe
    }

Conservation laws
    B(total) = B₁ + B₂ = 0
```

```
L(total) = L₁ + L₂ = 0
Q(total) = Q₁ + Q₂ = 0
```

3. Inflation Mechanism

```
Natural inflation from spin transition
a(t) = a₀exp(Ht)[1 + η∫Sμνλ(JσKμν)dΣλ]

Horizon evolution
dH = c/H[1 + f(preservation)]{
    Pre-transition: Causal connection
    Transition:     Geometric flip
    Post-transition: Expansion
}

Field dynamics
Φ(t) = Φ₀exp(iωt){
    ω = ω₀[1 + κ(ψ̄γμψ)Sμ]
    Phase coherence maintained
}
```

4. Observable Consequences

CMB signatures
$\delta T/T \approx 10^{-5}[1 + \text{preservation terms}]$

Power spectrum
$P(k) = P_0(k)[1 + \alpha(k/k_0)^{(1/3)}]\exp[-\beta(k/k^*)^2]$

Correlation functions
$\xi(r) = \xi_0(r)[1 + \gamma \int S\mu\nu\lambda(J\sigma K\mu\nu)d\Sigma\lambda]$

Specific predictions
1. Scale-dependent fluctuations
2. Geometric patterns in CMB
3. Preserved quantum correlations
4. Transition echoes

5. Testing Framework

Key observations
a) Light element abundances
b) CMB patterns
c) Large-scale correlations
d) Quantum signatures

Distinctive features
1. No initial singularity
2. Natural inflation
3. Preserved symmetries
4. Continuous transition
5. Observable patterns

This framework:

1. Resolves BBN constraints
2. Explains matter-antimatter asymmetry
3. Provides natural inflation
4. Maintains causality
5. Makes testable predictions

The early universe emerges as a natural consequence of:
- Spinor-mediated transitions
- Preservation hierarchy
- Balanced symmetries
- Geometric continuity

V. QUANTUM SCALE IMPLICATIONS

A. Fundamental Forces

The preservation hierarchy reveals how fundamental forces emerge and unify through spinor-mediated transitions.

1. Modified Gravity Strength

```
Complete gravitational coupling
    G_eff(r) = G₀[1 + κ(ψ̄γμψ)Sμνλ(Jσ/ρ)]{
        Quantum scale: G ∝ (M/mp)^(1/2)
        Middle scale:  G ∝ (M/mp)^(1/3)
        Cosmic scale:  G ∝ (M/mp)^(1/4)
    }

Force unification scale
    λunify = lp√(α_G/α_EM)[1 + f(preservation)]

Angular momentum contribution
    F_total = FNewton + κ∫Sμνλ(JσKμν)dΣλ
```

2. Strong Cp Resolution

```
Balanced CP violation across universes
    θQCD(total) = θ₁ + θ₂ = 0

Local CP parameter
    θ_local = θ₀[1 + κ(ψ̄γμψ)Sμ]{
        Universe 1: +θ
        Universe 2: -θ
    }

Preservation mechanism
    L_CP = κCP(ψ̄γ5γμψ)(Sλμν/ρc)exp(±iπ)
```

3. Force Unification

Unified coupling constant
  $\alpha\_unified(\lambda) = \alpha_0[1 + f(\lambda/\lambda p)]\{$
    EM: $\alpha\_EM(\lambda)$
    Weak: $\alpha\_W(\lambda)$
    Strong: $\alpha\_S(\lambda)$
    Gravity: $\alpha\_G(\lambda)$
  $\}$

Scale-dependent unification
  $g^2(\lambda) = g_0^2[1 + \eta \int S\mu\nu\lambda(J\sigma K\mu\nu)d\Sigma\lambda]$

Preservation hierarchy coupling
  $\kappa force(r) = \kappa_0 \exp(-r/\lambda)[1 + \text{preservation terms}]$

4. Quantum Interactions

Modified interaction vertices
  $\Gamma\mu\nu\lambda = \Gamma standard[1 + \kappa(\bar{\psi}\gamma\mu\psi)S\mu]$

Field strength tensors
  $F\mu\nu = \partial\mu A\nu - \partial\nu A\mu + [A\mu, A\nu] + S\mu\nu\lambda \nabla\lambda\Phi$

Preservation of quantum numbers
  $Q\_preserved = Q_0 \exp(i\omega t)[1 + f(\text{transition})]$

5. Observable Consequences

Specific predictions
  1. Force strength scaling:
     $\alpha(\lambda) = \alpha_0[1 + \text{scale-dependent terms}]$

  2. Unified behaviors:
     $g\_unify = g_0[1 + \text{preservation effects}]$

  3. Quantum correlations:

> $\langle\psi|\psi'\rangle = \exp(i\theta)[1 + \kappa(\bar{\psi}\gamma\mu\psi)S\mu]$
>
> 4. Transition signatures:
> $h(f) = h_0(f)[1 + \text{modification terms}]$

Key implications:

1. Gravity not fundamentally weak
2. Natural CP conservation
3. Smooth force unification
4. Preserved quantum properties
5. Observable signatures

The framework reveals:

- How forces naturally unify
- Why CP violation balances
- How gravity connects to other forces
- Where quantum effects manifest

B. Particle Physics

1. Spinor Dynamics

Complete spinor evolution
$$D\mu\psi = [\partial\mu + (1/4)\omega\mu ab\gamma a\gamma b + \kappa S\mu]\psi \{$$
  Phase I: Alignment (0° to 180°)
  Phase II: Transition (180° to 540°)
  Phase III: Completion (540° to 720°)
}

Preservation-modified Dirac equation
$$[i\gamma\mu(\partial\mu + \Gamma\mu) - m(1 + \kappa S\mu\bar{\psi}\gamma\mu\psi)]\psi = 0$$

Spin-torsion coupling
$$S\mu\nu\lambda = \eta(\bar{\psi}\gamma[\mu\gamma\nu\nabla\lambda]\psi)/\rho c$$

2. Cp Violation Mechanisms

Balanced CP violation
$$CP(\psi) \to \exp(i\theta CP)[1 + \kappa(\bar{\psi}\gamma 5\gamma\mu\psi)S\mu]\psi^*$$

Cross-universe symmetry
$$\theta CP(U_1) = -\theta CP(U_2)$$
CPT(total) = preserved

Generation mechanism
$$L\_CP = \kappa CP(\bar{\psi}\gamma 5\gamma\mu\psi)(S\lambda\mu\nu/\rho c)\{$$
  Matter universe: +CP violation
  Antimatter universe: -CP violation
}

3. Quantum Measurement

Preservation-mediated collapse
$|\psi\rangle \to |\psi'\rangle$ through:
  1. Spinor alignment with measurement basis
  2. 720° geometric transition

3. Information preservation

Measurement outcome
$P(\text{outcome}) = |\langle\psi|\psi'\rangle|^2[1 + \eta(\text{preservation terms})]$

Coherence maintenance
$\rho' = \rho + \kappa \int S\mu\nu\lambda(J\sigma K\mu\nu)d\Sigma\lambda$

C. Field Theory

1. Modified Quantum Field Theory

```
Action with preservation terms
  S = ∫d⁴x√-g{
    SQFT_standard +
    Spreservation +
    Scoupling
  }

Field equations
  δS/δφ = 0 yields:
  (□ + m² + κR)φ + λφ³ + η(ψ̄γμψ)Sμφ = 0

Propagator modification
  G(x,y) = G₀(x,y)[1 + f(preservation)]
```

2. Vacuum Energy

```
Modified vacuum expectation
  ⟨0|T^μν|0⟩ = T^μν_standard[1 + κ(ψ̄γμψ)Sμ]

Zero-point energy
  E_vac = E₀[1 + η∫Sμνλ(JσKμν)dΣλ]

Scale-dependent modification
  ρΛ(λ) = ρΛ_standard[1 + f(λ/λp)]
```

3. Particle Creation/Annihilation

```
Modified creation operators
  a†(k) → a†(k)[1 + κ(ψ̄γμψ)Sμ]

Pair production rate
```

$\Gamma(\text{pair}) = \Gamma_0 \exp(-2m/T)[1 + \text{preservation terms}]$

Vacuum polarization
$\Pi(q^2) = \Pi_0(q^2)[1 + \eta \int S\mu\nu\lambda(J\sigma K\mu\nu)d\Sigma\lambda]$

4. Observable Consequences

Key predictions
1. Quantum transitions:
   Modified probability amplitudes
   Preserved coherence patterns
   Distinctive interference effects

2. Vacuum effects:
   Scale-dependent energy density
   Modified Casimir force
   Quantum field correlations

3. Particle phenomena:
   Enhanced creation rates
   Preserved quantum numbers
   Specific decay patterns

This framework reveals:

1. Natural quantum measurement mechanism
2. Balanced CP violation
3. Modified vacuum structure
4. Preservation of quantum information
5. Observable quantum signatures

Key implications:

- Quantum-classical transition explained
- Vacuum energy naturally regulated
- Particle-antiparticle symmetry maintained
- Information preserved through transitions
- Testable quantum predictions

VI. OBSERVATIONAL PREDICTIONS

A. Gravitational Waves

In the Spinor-Mediated Universal Transition (SMUT) model, gravitational wave observations become critical for verifying the theory's predictions, particularly through distinctive **echo patterns**, **resonance frequencies**, and **detection strategies**. These elements offer measurable signatures in gravitational wave observatories and could provide insights into the layered structure of the universe as proposed by the preservation hierarchy.

1. Echo Patterns: Sequential Layered Transitions

The SMUT model predicts unique echo patterns due to the layered preservation hierarchy (classical, quantum, and conformal). As gravitational waves pass through regions of critical density or undergo transitions between these layers, they trigger sequential activations. This results in a cascade of echoes—secondary waves repeating the initial signal with delayed intervals.

- **Primary and Secondary Delays**: The echo pattern consists of a primary delay, directly associated with the **quantum-conformal layer transition**, and a secondary delay tied to the **conformal-classical layer transition**. These delays manifest as **echo cascades**, with the intensity and timing directly connected to the underlying transition layers.
- **Echo Cascade Effect**: This cascade effect results in a "stepping down" of wave intensity, producing a series of smaller echoes following the primary gravitational wave event. These echoes are anticipated to have predictable intervals, directly tied to the layer coupling constants (e.g.,

λ_1, λ_2, and λ_3 for classical, quantum, and conformal layers, respectively).

2. Resonance Frequencies

Resonance frequencies are another direct observable consequence of the SMUT model. As gravitational waves interact with the spinor-torsion structure at critical densities, resonance frequencies emerge, reflecting the coupling between the hierarchical layers.

- **Primary Frequency (f_{12})**: This frequency corresponds to the **quantum-conformal layer coupling**, scaling inversely with black hole mass as $f_{12} \propto M^{-1/4}$. This dependency offers a direct correlation between black hole mass and quantum-conformal layer interactions, highlighting quantum-scale effects on large cosmic structures.
- **Secondary Frequency (f_{23})**: Linked to the **conformal-classical layer coupling**, this frequency scales as $f_{23} \propto M^{-1/3}$. Observing this frequency would affirm the conformal geometry's role in regulating the classical gravitational behavior and allow distinctions between regions dominated by quantum coherence versus classical conservation laws.
- **Beat Frequency (f_beat)**: The combined effect of primary and secondary frequencies produces a beat pattern, calculated as $f_{beat} = |f_{12} - f_{23}|$. This beat frequency reflects the interference between the hierarchical transitions and provides a "fingerprint" of the layered structure predicted by the SMUT model.

3. Detection Strategies

To identify these distinctive features, gravitational wave observatories could implement specialized detection strategies focusing on the echo cascades and resonance frequencies described.

- **Echo Pattern Detection**: To capture the echo cascades, observatories should utilize **machine learning algorithms** designed to recognize repeating wave patterns with diminishing amplitudes at specific intervals. Training algorithms on simulated SMUT echo patterns could increase sensitivity and improve recognition of layered activations within gravitational wave signals.
- **Frequency Analysis for Resonance Signatures**: For resonance frequencies, **spectral analysis techniques** could be refined to identify the primary, secondary, and beat frequencies. Observatories may need to apply **Fourier Transform-based methods** to capture these oscillatory patterns, adjusting for mass-dependent scaling factors that would shift the frequencies based on the black hole or other massive structure involved.
- **Multi-Detector Coordination**: Given the need for high sensitivity, a coordinated effort across multiple detectors (e.g., LIGO, Virgo, and KAGRA) could enhance the signal-to-noise ratio. Cross-correlation of data from multiple observatories could also help verify the universal nature of the patterns, distinguishing them from local noise.

By focusing on these patterns and resonances, gravitational wave observatories can potentially detect and validate the unique gravitational wave signatures predicted by the SMUT model. If confirmed, these observations would not only support the SMUT framework but also provide evidence of a deeper layered structure to spacetime, fundamentally altering our understanding of gravitational wave behavior across cosmic scales.

B. Cosmic Structure

The SMUT model introduces novel mechanisms for **galaxy formation**, **large-scale patterns**, and **distribution functions** based on spinor-preserved angular momentum and the preservation hierarchy. Observing these patterns in cosmic structures would serve as further validation of the model, revealing a universe shaped by conserved rotational dynamics and seamless transitions across cosmic scales.

1. Galaxy Formation

In the SMUT model, galaxy formation is heavily influenced by **spinor-preserved angular momentum** and **torsion effects**. These effects create rotational dynamics that guide the formation and structure of galaxies, giving rise to unique characteristics observable at galactic scales.

- **Spiral Patterns and Angular Momentum Preservation**: The SMUT model predicts that galaxies will exhibit specific spiral patterns driven by preserved angular momentum within the preservation hierarchy. Galaxies are expected to show **consistent rotational velocities** that align with predictions of the spinor-torsion coupling mechanism rather than standard dark matter explanations.
- **Observable Rotation Curves**: Rather than attributing rotation curve anomalies to dark matter, the SMUT model proposes that these curves reflect **preservation effects** within the classical and quantum layers. Observing rotation curves that deviate from classical Newtonian predictions could provide evidence of the spinor-driven dynamics predicted by the model.
- **Twin Universe Influence on Galaxy Structure**: Galaxy formations may also show asymmetrical features or paired rotational alignments, as the model allows for the creation of "twin universe" effects. These would appear as **complementary structures** or paired angular momenta across distant galaxies, potentially detectable through large-scale surveys.

2. Large-Scale Patterns

Large-scale cosmic structures, such as the cosmic web, clusters, and voids, can reveal the effects of **scale-invariant patterns** generated by the preservation hierarchy and spinor-mediated transitions.

- **Fractal, Nested Spherical Geometry**: The SMUT model predicts a fractal structure in cosmic patterns, where clusters and voids repeat across different scales. This nested spherical symmetry results from the preservation of angular momentum across hierarchical layers. Observing fractal characteristics in the cosmic web—self-similar structures at varying scales—would support the model's prediction of geometric continuity across scales.
- **Spin-Driven Web Structure**: The cosmic web's filaments and nodes may display specific **rotational patterns** that reflect the SMUT model's hierarchical conservation of angular momentum. This would appear as large-scale alignment of angular momentum vectors or **preferred directions** in the distribution of galaxy clusters and filaments.
- **Scale-Dependent Distribution of Matter**: The preservation hierarchy suggests a scale-dependent distribution where **density fluctuations** become more pronounced as structures transition between hierarchical layers. Observing these transitions could reveal how matter distribution scales across cosmic distances and align with the SMUT model's layered geometry.

3. Distribution Functions

The SMUT model provides predictions for distribution functions related to mass, velocity, and spatial distribution that should be observable in cosmic surveys.

- **Mass Distribution Function**: The model suggests a **mass distribution function** influenced by preserved angular momentum, which could produce a characteristic curve differing from traditional models. Mass concentrations within galaxies and clusters may appear more uniform at specific scales, aligning with the conservation hierarchy. This could lead to mass distributions that fit a **power-law** function, reflecting the fractal nature of cosmic structures.

- **Velocity Dispersion Profile**: In galaxy clusters, the SMUT model predicts **velocity dispersion** that scales with cluster mass differently than standard dark matter models. The velocity profiles of galaxies within clusters should reflect **additional rotational dynamics** due to torsion and angular momentum preservation, producing distinct dispersion patterns detectable through spectroscopic surveys.
- **Spatial Correlation Function**: The spatial distribution of galaxies, quantified by the **two-point correlation function**, may show unique **correlation lengths** that correlate with the scale-dependent features of the SMUT model. The preservation hierarchy suggests that galaxies at larger separations might show weaker correlations than expected, with clustering features linked to resonance effects across hierarchical layers.

Observational Strategies:

1. **Cosmic Surveys and Fractal Analysis**: Using large-scale galaxy surveys (such as those from the Sloan Digital Sky Survey or Euclid) to analyze fractal patterns in the cosmic web could test the SMUT model's predictions. Identifying self-similar structures across multiple scales would support the presence of the nested spherical geometry predicted by the model.
2. **Rotation Curve Studies and Velocity Dispersion Analysis**: Observing rotation curves within galaxies and velocity dispersions in clusters can help distinguish SMUT model predictions from dark matter effects. Analyzing

these properties across galaxies of different masses and types could provide insights into the spinor-preserved dynamics.
3. **Correlation Functions and Statistical Distributions**: Using correlation functions and mass distributions from cosmic data, astronomers could identify scale-dependent features and distribution patterns that align with the SMUT model's hierarchical layers.

Through these observable features—galaxy formation, large-scale patterns, and specific distribution functions—the SMUT model offers a distinct lens through which cosmic structure can be analyzed. Validation of these predictions could offer a profound shift in our understanding of cosmic evolution, positioning the SMUT model as a viable framework that naturally accounts for galaxy formation and large-scale structure without invoking exotic matter.

C. High-Energy Phenomena

In the Spinor-Mediated Universal Transition (SMUT) model, high-energy phenomena such as cosmic rays, fast radio bursts (FRBs), and other novel astrophysical events arise as natural consequences of spinor-torsion interactions and transitions within the preservation hierarchy. These phenomena provide an important testing ground for the model, potentially revealing distinctive features tied to the layered structure of spacetime.

1. Cosmic Rays

Cosmic rays, especially at ultra-high energies, can be explained by the SMUT model as particles that gain additional energy through preservation transitions.

- **Energy Boost from Preservation Transitions**: When cosmic rays cross transition boundaries within the hierarchical layers (e.g., quantum to conformal layers), they experience an energy boost due to the spinor-torsion coupling. This results in cosmic rays with **extended high-energy tails** beyond standard predictions, which could be observable as a distinctive high-energy spectrum.

- **Directional Patterns and "Hotspots"**: The SMUT model also suggests that regions of high density, such as near black holes or neutron stars, could act as **cosmic ray acceleration sites** due to spinor-torsion interactions. This might lead to **directional anisotropies** or "hotspots" in cosmic ray distributions, where certain regions of the sky show elevated cosmic ray counts, offering a unique signature of the model's layered structure.

2. Fast Radio Bursts (Frbs)

In the SMUT framework, FRBs could result from small-scale transitions within the preservation hierarchy, particularly involving quantum or conformal layers.

- **Emission from Layered Transitions**: FRBs may be produced when high-density astrophysical objects, such as neutron stars, undergo **small-scale preservation transitions**. These events release a sudden, intense burst of radio energy as particles and fields realign through spinor-mediated interactions, potentially accounting for the short-duration yet high-energy nature of FRBs.
- **Polarization and Spinor Effects**: Due to the unique rotational symmetry of spinor fields (720-degree rotation), the SMUT model predicts that FRB signals may exhibit **specific polarization patterns**. Observed polarization could vary in a way that reveals underlying spinor-torsion dynamics rather than merely magnetic field effects, providing a potential observational signature of the SMUT model.
- **Repetition and Echo Patterns**: Some FRBs display repeating bursts, which the SMUT model interprets as echo effects from layer-specific transitions. This

repetition pattern may align with the echo cascade effects observed in gravitational wave predictions, suggesting that FRBs are a small-scale analog of these phenomena.

3. Novel Predictions

In addition to cosmic rays and FRBs, the SMUT model proposes several other high-energy phenomena tied to preservation transitions and spinor interactions:

- **Transition-Induced Gamma-Ray Bursts (GRBs)**: The model suggests that GRBs could result from major transitions within high-energy environments, such as during the collapse of massive stars or neutron star mergers. These **transition-driven bursts** would display layered emissions, where primary and secondary bursts reflect interactions across different layers, producing a characteristic profile observable in gamma-ray data.
- **High-Energy Neutrino Patterns**: Spinor-torsion interactions in the SMUT model could also influence neutrino emissions, resulting in **high-energy neutrinos** with unique directional distributions and energy profiles. Observatories like IceCube could detect these neutrinos as they carry distinct characteristics from preservation transitions, potentially aligning with cosmic ray and FRB data.
- **Magnetic Monopole Prohibition**: Notably, the SMUT model prohibits magnetic monopoles due to the preservation hierarchy's enforcement of [Equation] $\nabla \cdot B = 0$ across all layers. This consistency is maintained by the topological constraints of spinor rotations, which inherently prevent isolated magnetic poles from forming. This prediction is distinctive among theoretical models, setting the SMUT framework apart.

Observational Strategies:

1. **Cosmic Ray and Neutrino Observatories**: Observing extended energy spectra, directional hotspots, and unique high-energy neutrino profiles would offer direct evidence for SMUT model predictions. Instruments like the Pierre Auger Observatory and IceCube can focus on detecting any anomalies in energy distributions and directional patterns that align with preservation transition effects.
2. **FRB Polarization Studies**: Radio telescopes like CHIME and FAST could investigate FRB polarization for signatures of spinor dynamics, examining polarization shifts or rotations that might signal underlying preservation transitions.
3. **Gamma-Ray Burst (GRB) Analysis**: Observing GRBs with layered emission profiles could further support the SMUT model. Instruments sensitive to gamma-ray spectra can examine GRB light curves for primary and secondary emissions or echo-like structures indicative of layered preservation transitions.

Through these high-energy phenomena—cosmic rays, FRBs, GRBs, and neutrinos—the SMUT model provides a set of distinctive, testable predictions. Observations that align with these predictions would support the SMUT framework's unique approach to explaining cosmic events, reinforcing the model's vision of a universe structured by smooth transitions and spinor-preserved coherence across all scales.

VII. DISCUSSION AND FUTURE WORK

A. Experimental Tests

Testing the Spinor-Mediated Universal Transition (SMUT) model requires specific experimental strategies to detect the unique signatures of spinor-torsion dynamics and the preservation hierarchy. These experimental tests focus on gravitational waves, cosmic structure, and high-energy phenomena. Below, we outline the **current capabilities**, **future requirements**, and **key observations** necessary to validate the SMUT model's predictions.

1. Current Capabilities

Several current experimental facilities and methods are well-suited to test specific aspects of the SMUT model:

- **Gravitational Wave Observatories** (e.g., LIGO, Virgo, KAGRA): These detectors are capable of observing gravitational wave patterns, making them suitable for detecting **echo cascades** and **resonance frequencies** predicted by the SMUT model. Advanced signal processing techniques, including Fourier Transform and machine learning algorithms, can enhance their sensitivity to the layered echo patterns characteristic of the model.
- **Cosmic Ray Observatories** (e.g., Pierre Auger Observatory): High-energy cosmic ray facilities can analyze cosmic ray distributions and identify potential **energy boosts** from preservation transitions. They are also equipped to search for **directional anisotropies** or hotspots, which might reveal high-density regions where spinor-torsion interactions are most pronounced.
- **Fast Radio Burst (FRB) and Radio Polarization**

Observatories (e.g., CHIME, FAST): These radio telescopes are capable of detecting polarization signatures in FRBs. Analyzing the polarization patterns and repetition rates of FRBs can help identify spinor-induced effects and **echo-like structures** that align with SMUT model predictions.
- **Neutrino Detectors** (e.g., IceCube): Current neutrino observatories can detect high-energy neutrinos from cosmic sources. IceCube, for instance, can search for **unique directional patterns** in neutrino events, potentially providing evidence of spinor-torsion interactions within high-density regions.

2. Future Requirements

While existing facilities can capture some of the SMUT model's predictions, enhanced capabilities and new technologies are needed to rigorously validate the model:

- **Increased Sensitivity in Gravitational Wave Observatories**: To detect subtle resonance frequencies and echo cascades across cosmic scales, gravitational wave observatories may need increased sensitivity. Next-generation detectors, like the **Einstein Telescope** and **LISA** (Laser Interferometer Space Antenna), with enhanced low-frequency sensitivity, could capture the **layered structure of gravitational wave signals** more effectively.
- **Expanded Polarization and Frequency Coverage in Radio Telescopes**: Enhanced radio telescopes with broader frequency ranges and more precise polarization measurement capabilities could provide better insights into spinor-driven FRB phenomena. A telescope array specifically optimized to detect rapid polarization changes and echo patterns in FRBs would further validate SMUT model predictions.
- **Advanced Cosmic Ray and Neutrino Detectors**: For high-energy cosmic rays and neutrinos, larger detector arrays and more refined particle identification are required to detect the **energy distribution signatures**

of spinor transitions. Facilities that combine cosmic ray and neutrino detection, particularly in space-based observatories, could significantly expand detection ranges and provide a clearer view of the anisotropies and high-energy tails predicted by the SMUT model.
- **Cosmic Surveys with Higher Resolution and Scale Depth**: Large-scale galaxy surveys with finer resolution and deeper field coverage, such as those planned for **Euclid** and **LSST (Vera C. Rubin Observatory)**, would be valuable for observing **nested, fractal structures** and **rotational alignments** in cosmic distributions. These structures are predicted by the SMUT model's spinor-preserved angular momentum and self-similar patterns in the cosmic web.

3. Key Observations

The SMUT model suggests several key observations that would serve as experimental benchmarks for its validity:

- **Gravitational Wave Echoes and Resonances**: Detecting a **cascade of gravitational wave echoes** with primary and secondary delays would be a definitive test of the layered preservation hierarchy. Additionally, observing the **primary, secondary, and beat resonance frequencies** associated with specific mass scalings in black holes or other massive structures would support the model's predictions of layer-specific gravitational wave behaviors.
- **Directional Anisotropies in Cosmic Rays**: Observing directional "hotspots" or energy distribution anomalies in ultra-high-energy cosmic rays would validate the model's prediction that spinor-torsion interactions create **localized acceleration regions**. Such hotspots would likely appear near high-density regions, offering direct evidence of SMUT dynamics.
- **Polarization and Repetition Patterns in FRBs**: Analyzing FRB signals for distinct polarization patterns and repetition intervals could reveal spinor-induced effects. Specifically, the detection of **echo-like bursts**

with changing polarization would provide evidence for small-scale preservation transitions, which are a central component of the SMUT model.
- **Self-Similar Cosmic Structures and Galaxy Spin Alignments**: Identifying **fractal, nested spherical patterns** in the cosmic web, as well as **large-scale spin alignments** of galaxies and clusters, would support the SMUT model's prediction of a universe organized by self-similar, spinor-preserved structures. These structures reflect the conservation of angular momentum across hierarchical layers, a unique aspect of the SMUT framework.
- **Neutrino Distribution Patterns**: Observing high-energy neutrinos with specific directional correlations or unusual energy profiles would suggest the influence of spinor-torsion dynamics. This could align with cosmic ray observations and provide an independent confirmation of high-energy preservation transitions.

Together, these **experimental tests** would provide a robust approach to validating or refuting the SMUT model. Current technology provides a strong foundation, but achieving definitive results will require **next-generation observatories** and **targeted detection strategies**. Confirming these key observations would fundamentally shift our understanding of cosmic structure, high-energy phenomena, and gravitational wave behaviors, lending significant support to the SMUT model as a unified theory of cosmic and quantum phenomena.

B. Theoretical Extensions

The Spinor-Mediated Universal Transition (SMUT) model offers opportunities to expand theoretical understanding across several key areas: **beyond the Standard Model physics**, **quantum gravity**, and **cosmological evolution**. These extensions leverage the preservation hierarchy and spinor-torsion dynamics to provide fresh insights into longstanding theoretical challenges and open up new directions for research.

1. Beyond The Standard Model

The SMUT model proposes that many phenomena traditionally attributed to unknown particles or forces can instead be explained through preservation dynamics and spinor-torsion interactions, offering alternatives to some **beyond Standard Model (BSM) theories**:

- **Reinterpreting Dark Matter**: Rather than hypothesizing exotic dark matter particles, the SMUT model suggests that what appears as dark matter effects in galactic rotation curves and gravitational lensing may actually be due to **preserved angular momentum and torsion effects** in the layered preservation hierarchy. This shift could eliminate the need for dark matter particles and instead attribute observed anomalies to **spinor-preserved structures** within cosmic scales.
- **Absence of Magnetic Monopoles**: The model's prohibition of magnetic monopoles through topological constraints and flux conservation ([Equation] $\nabla \cdot B = 0$) is a unique stance that provides a natural explanation for their absence in experiments. This suggests that magnetic monopoles are incompatible with a universe organized by preservation layers, distinguishing the SMUT model from BSM theories that allow for monopoles.

- **Matter-Antimatter Asymmetry and Baryogenesis**: The SMUT model suggests that **paired universe production** and spinor-mediated CP violation within hierarchical layers could resolve the matter-antimatter asymmetry. Rather than relying on additional fields or particles, this approach attributes baryogenesis to **spinor symmetry-breaking** in paired universes, providing an alternative to conventional mechanisms like leptogenesis or supersymmetric particles.

2. Quantum Gravity Implications

The SMUT model presents a framework that naturally bridges quantum mechanics and general relativity through **spinor-torsion interactions** and **layered preservation**, addressing key quantum gravity challenges without the need for additional dimensions or discrete spacetime:

- **Spinor-Torsion as Quantum-Gravitational Mediators**: The model introduces **spinor fields with torsion** as intrinsic components of spacetime at high densities, proposing that spinor fields mediate quantum-gravitational effects. This could lead to a quantized approach to gravity that maintains smooth, continuous spacetime transitions rather than relying on the spacetime discreteness suggested by loop quantum gravity or string theory.
- **Avoidance of Singularities**: By incorporating a layered preservation hierarchy, the SMUT model offers a **natural mechanism to avoid singularities** in black holes and the early universe. Spinor-torsion coupling allows spacetime to adjust dynamically at critical densities, leading to **smooth transitions at high-curvature regions** without the need for infinite densities. This could resolve the black hole information paradox by maintaining information continuity across these transitions.
- **Unified Framework for Quantum-Classical Transitions**: The SMUT model provides a coherent

theory for the quantum-to-classical transition through spinor alignment and 720° rotational symmetry, offering a quantum gravity framework that naturally explains **wavefunction collapse as a preservation transition**. This interpretation of quantum measurement within a gravitational context provides an elegant solution to the measurement problem and aligns quantum coherence with classical causality.

3. Cosmological Evolution

The SMUT model's layered, spinor-mediated structure provides a foundation for a **self-regulating, cyclic universe** that can address longstanding cosmological issues:

- **Big Bang and Big Spin as Initial Conditions**: The SMUT model replaces the Big Bang singularity with a **Big Spin transition**, where the initial universe undergoes a spinor-mediated transition, generating preserved angular momentum that drives cosmic expansion. This initial "Big Spin" provides a natural explanation for large-scale cosmic rotation patterns and the observed isotropy of the cosmic microwave background (CMB).
- **Dark Energy as a Consequence of Preservation Dynamics**: Instead of introducing an external dark energy field, the SMUT model interprets dark energy as a natural outcome of **preservation hierarchy dynamics**. The layered preservation structure introduces a scaling effect on spacetime, producing the observed acceleration in cosmic expansion without requiring an additional force, allowing dark energy to emerge as a structural property of the universe.
- **Fractal, Scale-Invariant Cosmic Structure**: The SMUT model's hierarchical preservation layers lead to **self-similar, fractal cosmic structures**, with galactic and cluster-scale patterns that repeat across scales. This fractal organization, arising naturally from the preservation of spinor-driven angular momentum, suggests that cosmic

evolution is governed by a scale-invariant structure, providing a testable prediction for large-scale structure surveys.

- **Potential for a Cyclic Universe**: With smooth transitions and spinor-mediated information preservation, the SMUT model supports a **cyclic cosmological model** where the universe undergoes continuous cycles of expansion, preservation, and regeneration. This cyclic behavior would resolve the thermodynamic constraints of a single, linear universe model, suggesting that the universe might continuously evolve while maintaining continuity across cycles.

These theoretical extensions demonstrate the SMUT model's potential to reshape our understanding of fundamental physics, offering new approaches to quantum gravity, cosmic evolution, and beyond Standard Model phenomena. If validated, this framework could unify quantum mechanics and relativity within a layered, spinor-driven universe, providing a seamless and self-consistent approach to understanding both microscopic and cosmic scales

C. Technological Applications

The Spinor-Mediated Universal Transition (SMUT) model opens up novel avenues for technological applications, especially in areas like **detection methods**, **instrument development**, and **advanced data analysis**. These technologies aim to enhance our ability to observe and interpret the unique phenomena predicted by the SMUT model, providing new ways to explore the layered structure of spacetime and spinor-torsion dynamics.

1. Detection Methods

To capture the subtle signals and transitions predicted by

the SMUT model, new detection methods can be tailored for gravitational waves, cosmic rays, and high-energy astrophysical phenomena.

- **Gravitational Wave Echo Detection**: The SMUT model predicts gravitational wave echoes resulting from layer transitions within the preservation hierarchy. **Machine learning algorithms** designed to recognize echo patterns with specific delay intervals could improve sensitivity in gravitational wave observatories. These algorithms could be trained on synthetic SMUT signals to identify primary and secondary delays and resonance frequencies in real data, improving the detection of gravitational wave cascades.
- **Polarization Measurement for FRBs**: Given that the SMUT model predicts unique polarization patterns in fast radio bursts (FRBs) due to spinor effects, new methods focused on **real-time polarization analysis** would be beneficial. Enhanced polarization-sensitive radio receivers and processing systems could capture rapid changes and angular distributions of polarization, providing insight into spinor-torsion interactions at astrophysical scales.
- **Cosmic Ray and Neutrino Angular Pattern Analysis**: Since SMUT dynamics might generate directional anisotropies, new detection methods that analyze **angular distributions** in cosmic rays and neutrinos would be valuable. Direction-sensitive detectors could measure incoming angles with precision, helping researchers identify cosmic regions associated with spinor-preserved angular momentum and potentially verifying the SMUT model's predictions.

2. New Instruments

New instruments tailored to the specific requirements of the SMUT model's predictions would provide better data on gravitational waves, FRBs, cosmic rays, and large-scale structure patterns:

- **Next-Generation Gravitational Wave Detectors**: Enhanced gravitational wave observatories such as the **Einstein Telescope** and **LISA** (Laser Interferometer Space Antenna) are critical for observing the resonance frequencies and layered echoes predicted by the SMUT model. These detectors will offer better low-frequency sensitivity and longer baselines, allowing for the detection of the subtle wave patterns arising from transitions between preservation hierarchy layers.
- **Polarization-Sensitive Radio Arrays**: New radio arrays designed with advanced polarization sensitivity, such as the planned **Square Kilometre Array (SKA)**, could observe the predicted spinor-torsion effects in FRBs. Arrays optimized for multi-channel polarization data would allow scientists to track polarization fluctuations and correlate these with spinor field dynamics, as predicted by the SMUT model.
- **High-Energy Particle and Neutrino Observatories**: To detect high-energy cosmic rays and neutrinos with the energy profiles and angular distributions described in the SMUT framework, new particle detectors with improved **energy resolution and angular tracking** capabilities are needed. **Next-generation space-based observatories** could capture cosmic rays and neutrinos over larger volumes, detecting the high-energy tails and directional patterns unique to SMUT dynamics, while ground-based neutrino telescopes could verify high-energy events.
- **Deep-Sky Surveys with High Angular Resolution**: For cosmic structures, instruments such as the **Vera C. Rubin Observatory (LSST)** and the **Euclid** mission can provide high-resolution, deep-sky surveys capable of observing the scale-invariant, fractal patterns predicted by the SMUT model. These surveys would support tests of large-scale structure predictions and help validate the SMUT model's view of a self-similar cosmic web.

3. Advanced Data Analysis

To interpret the complex data from these advanced instruments and detection methods, the SMUT model's predictions require new approaches in data processing and analysis:

- **Multi-Scale Analysis Algorithms**: The SMUT model predicts self-similar structures across cosmic scales. Algorithms capable of **multi-scale fractal analysis** would allow researchers to detect nested, scale-invariant patterns in cosmic data. Such algorithms could be applied to galaxy distributions, filament structures, and rotation patterns to identify the fractal, nested geometry characteristic of SMUT's cosmic structure.
- **Fourier Transform and Resonance Analysis in Gravitational Waves**: The resonance frequencies predicted by the SMUT model require precise frequency analysis tools. Enhanced **Fourier Transform techniques** and **resonance frequency identification** algorithms could isolate the primary, secondary, and beat frequencies in gravitational wave data, revealing the layered transitions and beat patterns that align with SMUT theory. Applying these techniques to data from advanced gravitational wave detectors would help isolate SMUT-specific signatures.
- **Polarization Data Interpretation with Machine Learning**: Machine learning models could analyze polarization data from FRBs to detect subtle patterns related to spinor rotation effects. These models could be trained to recognize polarization rotations, angular shifts, and time-correlated patterns, which would be indicative of spinor field dynamics within the SMUT framework.
- **Integrated Multi-Observatory Data Analysis**: The SMUT model's predictions span multiple observational domains. An integrated analysis framework combining data from gravitational wave detectors, cosmic ray observatories,

and neutrino telescopes could cross-reference signals and establish correlations across high-energy phenomena, revealing spinor-preserved structures. Cross-observatory data correlation would allow for a more comprehensive test of SMUT predictions.

These technological advancements—improved detection methods, specialized instruments, and sophisticated data analysis techniques—are essential for testing and validating the SMUT model's unique predictions. They could enable scientists to capture the layered, spinor-mediated structure of the universe and reveal new insights into cosmic and quantum phenomena, driving the development of unified theories in cosmology and quantum gravity.

VIII. CONCLUSION

The Spinor-Mediated Universal Transition (SMUT) model represents a significant advancement in the quest for a unified theory of the universe, bridging quantum mechanics and general relativity through a layered, spinor-torsion framework. By re-envisioning cosmic and quantum phenomena as expressions of a preservation hierarchy, the SMUT model offers a self-consistent explanation for a wide range of mysteries—from gravitational wave echoes and cosmic structure to the absence of magnetic monopoles and the emergence of the arrow of time.

Unification Achievement

Bridging Scales With A Preservation Hierarchy:

The SMUT model's layered structure, organized by classical, quantum, and conformal layers, establishes a natural continuity across scales. This layered approach enables a seamless connection between quantum behaviors and large-scale cosmic structures, maintaining coherence and conservation laws throughout the universe. By rooting gravitational, electromagnetic, and quantum phenomena within a single hierarchy, the SMUT model achieves a powerful unification of scales that has eluded traditional theories.

A New Understanding Of Cosmic And Quantum Phenomena:

Through the spinor-torsion mechanism, the SMUT model provides a unique explanation for observed galactic rotations, high-energy cosmic rays, FRBs, and gamma-ray bursts—all without invoking dark matter or extra dimensions. The model's innovative spinor-preserved angular momentum and torsion effects offer explanations for the formation of cosmic structures, the fractal organization of the universe, and the acceleration of cosmic expansion, all unified by the same underlying principles.

Resolution Of Long-Standing Paradoxes:

The SMUT model resolves several key challenges in physics, including the black hole information paradox, matter-antimatter asymmetry, and the absence of singularities. By maintaining a smooth transition mechanism at high densities, the SMUT model avoids singularities, ensures information continuity, and provides a natural arrow of time. This approach not only solves specific issues but suggests a broader principle of conservation and coherence that operates at every level of reality.

Implications For Quantum Gravity And The Nature Of Time:

By incorporating spinor fields with torsion as mediators of quantum-gravitational effects, the SMUT model moves closer to a true theory of quantum gravity, avoiding the need for discrete spacetime or additional dimensions. Time itself emerges as a layered, asymmetrical progression governed by the preservation hierarchy, offering a novel perspective on temporal flow and entropy. This layered approach to time challenges traditional notions and opens pathways for further exploration into the nature of causality and continuity.

Major Resolutions

The Spinor-Mediated Universal Transition (SMUT) model offers solutions to several longstanding challenges in physics, providing a unified, spinor-torsion framework that addresses issues spanning cosmology, quantum mechanics, and gravitational theory. These resolutions demonstrate the model's capacity to maintain continuity across scales, from subatomic particles to cosmic structures, and to offer fresh perspectives on phenomena traditionally explained by exotic particles or forces.

1. Resolution Of Singularities And Black Hole Information Paradox

The SMUT model's preservation hierarchy and spinor-torsion dynamics eliminate the need for singularities, providing smooth, continuous transitions at high densities. Unlike traditional models, which predict infinite densities at black hole cores, the SMUT framework enables:

- **Singularity Avoidance**: Spinor fields with torsion introduce a "twist" in spacetime geometry that maintains continuity at critical densities, preventing singularities and ensuring a finite, structured internal geometry within black holes.
- **Information Preservation**: The SMUT model preserves information across all layers through spinor-torsion interactions, addressing the black hole information paradox by ensuring that information is not lost but rather conserved in the transition hierarchy. This continuity offers a natural explanation for information retention in black hole dynamics.

2. Dark Matter Reinterpretation

The SMUT model eliminates the need for hypothetical dark matter particles by attributing galactic rotation anomalies and gravitational lensing effects to spinor-preserved angular momentum within the preservation hierarchy:

- **Galactic Rotation Curves**: The preservation of angular momentum across layers, particularly within the quantum and classical layers, provides the rotational dynamics observed in galaxies. This structure produces effects traditionally attributed to dark matter, explaining rotation curves without requiring additional matter.
- **Lensing Effects**: Spinor-torsion effects in high-density regions can influence gravitational lensing patterns, creating apparent mass where none exists. The SMUT model's layered structure thus accounts for lensing anomalies by modifying how mass and space are perceived across layers, offering a structural alternative to dark matter.

3. Absence Of Magnetic Monopoles

The SMUT model inherently prohibits the existence of magnetic monopoles due to its topological constraints and the conservation of magnetic flux across all preservation layers:

- **Flux Conservation**: In the SMUT framework, [Equation] $\nabla \cdot B = 0$ is preserved universally due to the spinor fields' topological structure, which maintains closed-loop magnetic field lines without isolated poles.
- **Spinor-Torsion Topology**: The unique 720° rotation symmetry of spinor fields creates a field structure that prevents the formation of magnetic monopoles. This prohibition aligns with experimental observations and provides a theoretical foundation for their absence in nature.

4. Matter-Antimatter Asymmetry And Paired Universes

The SMUT model offers a natural explanation for the matter-antimatter imbalance observed in the universe, linking it to a paired universe production mechanism within the preservation hierarchy:

- **Paired Universe Production**: The model suggests that the universe's matter-antimatter asymmetry arises from a balanced creation process in paired universes, with each universe containing equal but opposite matter distributions.
- **Spinor-Mediated CP Violation**: Within this framework, spinor-torsion interactions provide the necessary CP violations that result in observable matter dominance, while maintaining overall symmetry across universes. This approach eliminates the need for additional fields or particles, grounding the asymmetry in spinor dynamics.

5. Emergence Of Dark Energy And Cosmic Acceleration

The SMUT model proposes that dark energy, responsible for cosmic acceleration, is an emergent property of the layered preservation hierarchy, resulting from interactions within the conformal layer:

- **Intrinsic Layer Dynamics**: The conformal layer's geometric continuity introduces a scaling effect on spacetime that drives the expansion of the universe, aligning with observations of dark energy. This approach offers a structural explanation for cosmic acceleration without invoking an external dark energy field.
- **Unified Expansion Mechanism**: The layered nature of the SMUT model suggests that cosmic expansion is a product of the preservation hierarchy itself, with each layer influencing the observed rate of expansion. This self-sustaining structure provides a new perspective on dark energy as an inherent aspect of the universe.

6. Natural Emergence Of The Arrow Of Time

The SMUT model explains the arrow of time as an emergent property of sequential activations within the preservation hierarchy:

- **Layered Time Asymmetry**: Time asymmetry arises naturally from the order in which preservation layers activate, with the classical layer providing an apparent forward direction for time, while quantum and conformal layers contribute reversible and geometric continuity, respectively.
- **Entropy and Preservation**: The model's structure aligns with the Second Law of Thermodynamics, where entropy increase is a consequence of preservation transitions. This layered approach to time and entropy provides a fundamental explanation for the observed arrow of time.

7. Self-Similar Cosmic Structure And Scale-Invariance

The SMUT model predicts a self-similar, fractal structure for cosmic phenomena, guided by angular momentum conservation within the preservation hierarchy:

- **Fractal and Scale-Invariant Patterns**: The nested, fractal geometry predicted by the SMUT model explains the large-scale organization of the cosmic web and the scale-invariant distribution of matter.
- **Rotational Alignments and Spinor Effects**: Spinor-preserved angular momentum results in the alignment of cosmic spins and the formation of cosmic filaments, aligning with observations of structure formation without requiring additional forces or particles.

These resolutions highlight the SMUT model's potential to unify physics by providing a single, cohesive framework for addressing fundamental questions across disciplines. By grounding these phenomena in spinor-torsion dynamics and a layered preservation hierarchy, the SMUT model offers an elegant, self-consistent approach to the universe's most profound mysteries, fundamentally reshaping our understanding of cosmic and quantum realities.

Future Directions

The Spinor-Mediated Universal Transition (SMUT) model provides a robust framework that unites quantum mechanics and general relativity while explaining cosmic phenomena in terms of layered preservation. Moving forward, several promising directions could help expand, refine, and validate the SMUT model through theoretical advancements, empirical tests, and technological innovation.

1. Experimental Validation And Observation

Testing the SMUT model's predictions across gravitational wave signals, high-energy cosmic rays, and large-scale cosmic structures is essential to validate its claims.

- **Gravitational Wave Echoes and Resonances**: Observatories like LIGO, Virgo, and future detectors such as the Einstein Telescope and LISA could focus on detecting gravitational wave echo patterns and resonance frequencies. These signals would provide evidence for the preservation hierarchy's layered structure, including primary and secondary delays, unique to SMUT's framework.
- **Polarization and Repetition in FRBs**: Analyzing fast radio burst (FRB) polarization patterns and potential echo-like repetitions could reveal underlying spinor-torsion interactions. With advanced polarization-sensitive radio telescopes like the Square Kilometre Array (SKA),

researchers can look for spinor-specific polarization shifts that match SMUT model predictions.
- **Large-Scale Structure Surveys**: Next-generation cosmic surveys (e.g., the Vera C. Rubin Observatory and Euclid) can help detect the scale-invariant, fractal structures predicted by SMUT, especially in galaxy rotation alignments and the cosmic web. Observing such patterns across different scales would support the model's self-similar, preservation-driven structure.

2. Quantum Gravity Research

The SMUT model's spinor-torsion framework provides a new approach to quantum gravity, presenting further research opportunities to formalize and expand upon its gravitational implications.

- **Spinor Field Dynamics in Quantum Field Theory**: Developing a more detailed understanding of how spinor fields interact with torsion within quantum field theory could help bridge gaps between SMUT's theoretical foundation and traditional quantum field frameworks. This includes exploring how spinor fields mediate gravitational interactions at microscopic scales.
- **Unified Quantum-Classical Transition Mechanisms**: Research into how the SMUT model's layered transitions influence wavefunction collapse could offer a pathway for reconciling quantum mechanics and classical causality. By formalizing quantum measurement within the context of layered preservation, SMUT could provide a more coherent theory for the quantum-classical boundary.
- **Mathematical Consistency with Existing Quantum Gravity Theories**: Further work to align SMUT's predictions with other quantum gravity theories, like loop quantum gravity or holography, would help clarify where SMUT stands within the broader landscape. This comparison may highlight complementary aspects or novel distinctions that contribute to a more unified

understanding of gravity at quantum scales.

3. Cosmological Evolution And Cyclic Universe Models

SMUT's self-sustaining, cyclic nature suggests intriguing directions for exploring cosmic evolution beyond the standard Big Bang model.

- **The "Big Spin" as an Alternative to the Big Bang**: By replacing the Big Bang singularity with a structured "Big Spin" transition, the SMUT model posits an alternative origin of cosmic rotation patterns and isotropy. Future research could refine this idea to explore how spinor dynamics account for initial conditions of the universe, isotropy, and the CMB's alignment patterns.
- **Exploring a Cyclic Universe Model**: The preservation hierarchy suggests the potential for a cyclic universe, where expansion, preservation, and regeneration occur in continuous cycles. Examining how this cyclic model aligns with thermodynamic laws, entropy increase, and large-scale cosmic observations would help determine if SMUT can support a regenerative universe without the need for additional dimensions or forces.
- **Resolving Cosmological Parameters Through the Preservation Hierarchy**: Further research could focus on how SMUT's layered structure influences fundamental cosmological parameters, such as the Hubble constant, baryon-to-photon ratio, and dark energy density. This approach might reveal natural explanations for these parameters, potentially explaining cosmic expansion rates and acceleration.

4. Extensions Beyond The Standard Model Of Physics

The SMUT model's spinor-based framework opens the door to exploring Beyond Standard Model (BSM) physics, offering a structural alternative to dark matter and addressing unresolved questions in particle physics.

- **Reinterpreting Dark Matter and Dark Energy**: Future research can formalize SMUT's reinterpretation of dark matter as an effect of preserved angular momentum and torsion. Researchers could analyze cosmic data to further differentiate between predictions for dark matter particles and SMUT's structure-driven alternative, particularly in galactic rotation and cluster dynamics.
- **Matter-Antimatter Asymmetry and Paired Universes**: The SMUT model's approach to baryon asymmetry via paired universe production and CP violation through spinor-torsion interactions could provide insights into particle physics. Expanding on this paired universe concept may offer explanations for why matter dominates in our observable universe while maintaining a conserved symmetry across paired universes.
- **Prohibition of Magnetic Monopoles and Topological Constraints**: The SMUT model's inherent prohibition of magnetic monopoles through its topological and spinor rotation constraints offers an intriguing avenue for research. Exploring how this prohibition aligns with topological field theories and gauge symmetries could yield new insights into particle properties and conservation laws.

5. Data Analysis Innovations And Machine Learning Applications

To fully interpret SMUT's predictions, data from cosmic and quantum phenomena need advanced analytical methods, including machine learning and multi-scale analysis.

- **Machine Learning for Gravitational Wave and FRB Data**: Applying machine learning algorithms to detect echo patterns and polarization signatures in gravitational waves and FRBs would help isolate SMUT-specific signals. Algorithms trained on simulated SMUT model data could

identify subtle variations that traditional methods might overlook, particularly in low-amplitude or noisy signals.
- **Multi-Scale Fractal Analysis of Cosmic Structures**: To test SMUT's prediction of a fractal cosmic structure, multi-scale analysis algorithms are needed to detect nested, self-similar patterns across different scales. Such algorithms could apply to galaxy surveys, mapping fractal properties within the cosmic web and identifying rotational alignments unique to SMUT's layered structure.
- **Integrated Observational Data Analysis**: Given the SMUT model's predictions across gravitational waves, high-energy cosmic rays, and cosmic structure, an integrated data analysis framework could cross-reference signals from various observatories. Cross-correlation of data across domains could provide a comprehensive view of SMUT phenomena, revealing shared features that link cosmic structure, quantum gravity, and high-energy events.

Summary

These future directions offer a roadmap to expanding and testing the SMUT model, positioning it as a comprehensive framework that could potentially unify physics. With experimental advancements, theoretical exploration, and data analysis innovations, the SMUT model could reshape our understanding of cosmic structure, quantum gravity, and fundamental physical laws, bridging scales and domains through a unified, spinor-mediated framework.

A THOUGHTFUL CONSIDERATION

The Spinor-Mediated Universal Transition (SMUT) model provides a compelling vision of a universe in which continuity, structure, and preservation emerge naturally through layered spinor-torsion interactions. This model challenges traditional assumptions, offering a unified framework that elegantly connects phenomena across the quantum, classical, and cosmic scales. At its core, the SMUT model is a call to rethink foundational principles, suggesting that the universe's complexity can be understood through an interplay of layered transitions and conserved dynamics.

Reframing Foundational Concepts

One of the SMUT model's most profound contributions is its ability to reframe longstanding scientific concepts. By introducing the preservation hierarchy, it suggests that singularities, dark matter, and even time itself are emergent properties rather than fundamental constructs. This layered approach brings a new clarity to paradoxes that have long challenged both quantum mechanics and general relativity, inviting us to see them as part of a larger, interconnected whole.

Harmony Between Quantum and Cosmic Realms

The SMUT model's spinor-torsion framework creates a harmony between the quantum and cosmic realms, unifying seemingly disparate scales through a single, structured hierarchy. This approach challenges us to think beyond particle-based explanations for phenomena like dark matter and to consider instead a universe organized by intrinsic geometric and rotational principles. It is an elegant proposal that conservation and coherence are not just mathematical constructs but fundamental aspects of reality itself.

Inspiring a New Era of Exploration

By introducing testable predictions—such as gravitational wave echoes, FRB polarization shifts, and fractal cosmic structures—the SMUT model paves the way for a new era of scientific inquiry. Its success depends not only on theoretical elegance but also on the capacity of our instruments and methodologies to detect the subtle signatures of a spinor-driven universe. As we refine our detection methods, develop new instruments, and innovate in data analysis, we are equipped to delve into this layered reality and explore its implications.

An Invitation to Collaborative Discovery

Ultimately, the SMUT model is an invitation to collaborative discovery. It offers physicists, astronomers, and theorists a shared framework within which to explore foundational questions about existence, structure, and the nature of time. In its simplicity and coherence, the SMUT model reminds us that scientific progress often arises not from complication but from finding unity in diversity.

This model's elegant and innovative approach suggests that we may be on the brink of a paradigm shift, one that transforms our understanding of the universe and reveals a cosmos governed by preservation, coherence, and self-sustaining dynamics. The SMUT model encourages a thoughtful consideration of the universe, where every layer, every transition, and every phenomenon resonates with the underlying principles of a unified, interconnected reality.

www.ingramcontent.com/pod-product-compliance
Lightning Source LLC
Chambersburg PA
CBHW070240220526
45465CB00004B/1470